For Marj x

Hen Party

A Celebration

Arthur Parkinson

PARTICULAR BOOKS

CONTENTS

Preface .... *vii*
Ancona 2
Andalusian 4
Appenzeller Spitzhauben 6
Araucana 10
Barnevelder 12
Belgian Bantam 14
Brahma 18
Burford Brown 22
Cochin 24
Commercial layers 28
Cornish Game 32
Croad Langshan 34
Deathlayer 36
Derbyshire Redcap 38
Dorking 40
Dutch Bantam 44
Faverolles 46
Fayoumi 48
Friesian 50
Goldtop 52
Groninger Meeuwen 54
Hamburg 56
Hedemora 58
Houdan 60
Japanese Bantam 62
Jungle Fowl 64
Legbar 66
Leghorn 70
Lincolnshire Buff 74
Marans 76
Marsh Daisy 80

New Hampshire Red 82
Norfolk Grey 84
Old English Game 86
Old English Pheasant
   Fowl 90
Orpington 92
Pekin Bantam 96
Penedesenca 100
Plymouth Rock 102
Poland 104
Rhode Island Red 108
Rosecomb 110
Scots Dumpy 112
Scots Grey 116
Sebright 118
Sicilian Buttercup 122
Silkie 124
Silverudd 128
Spanish 130
Sultan 132
Sumatra 134
Sussex 136
Swedish Flower 140
Thuringian 142
Vorwerk 144
Welsummer 146
Wyandotte 148
Keeping Hens *153*
Glossary *171*
A Chart of Chicken Colours ... *176*
Other Resources *178*
Acknowledgements *179*

YOU GET VERY adept at combing your fingers through feathers. I've got a Buff Cochin hen on my lap who is a giant, billowing, cushion-like creation of a chicken. I'm going through her huge amount of tangerine-coloured plumage to dust her undersides, gently and fleetingly revealing her pink skin within the excess of fluff to puff her with a white mite-repelling powder. She clucks softly with chattering concern, and I mutter back to her in agreement. This task is more of a precaution than a need, but there is lot of what I shall term as 'feather dressing' for those of us who really care for and know our chickens. Such rituals are regular occurrences if our hens are to have a harmonious party!

This particular hen's salon appointment completed, I place her down onto the ground and she returns to the flock to little acknowledgement from the other hens.

The cockerel, however, has been spying on the whole process and delightfully clucks at her reappearance. His neck feathers frill out and rise all in unison, like an orange sea anemone crossed with a Lyles Golden Treacle tin lion.

He is a griffin, albeit one who is wearing feathered pyjamas. When you have rather a lot of hens and are breeding them, you can't really give them all names, but the cockerel is called John. He is this year's chosen cockerel and has been allowed to harem the hens due to his fine pedigree.

I keep my chickens on rented farm land. This is a long way from my childhood days of garden poultry-rearing in Nottingham. Wooden posts support fencing, surrounded with electric wire to keep my dear ones as safe as I can. Within this generous hen run where grass is worshipped I have a little shed as a store for feed, bales of bedding and endless poultry sundries. I sit in here when it rains hard, carefully finding a spot among the egg boxes that come and go in their number depending on the quantity of precious eggs being laid. For me this is a time of hard-won, reclusive peace, away from the madding crowd. The hens jolly me along with their company and their requirements and we are, I hope, a happy band. Who and what they think I am, I'll never know.

You only see chickens' characters emerge if you take the time each day to admire, observe and talk to them. The first thing you will notice is that hens do not live together quietly in communes of nun-like sisterhood. They are more like harlots with sharp-beaked personalities, and their aptly titled society, the pecking order, can ruffle into chaos at a moment's notice.

We have the older, plump, cock-trodden, red-faced madams grumpily hustling among their blossoming and demure daughters of the season. Then there are the smaller bantam hens who are either completely absorbed by their chicks, sitting quietly in their own private quarters on a clutch of eggs, or bustling about everyone else with a noticeably bad temper if there are no eggs for them to sit on. Such brattish attitudes ruffle the feathers of the other high-ranking hens and so the little bantams are chased about, resulting in more furious cackling.

You learn, essentially, to read what is normal behaviour. Sluggishness is always a warning that something is not well; the usual healthy decorum of the flock is one of a lively, alert and receptive band of interested minds, regarding the world about them.

For this hen party to occur at all, chickens must be tended to with consideration and care. They are creatures of habit who must be kept clean, fed well, given space and protected from predators – and even, at times, from one another.

How to keep them content requires the education of many books. This book, however, is primarily about their stunning diversity and the wealth of chicken characters to be found within the spirits of these pure and rare breeds that are becoming very rare indeed.

It might come as a surprise to hear that chickens are endangered, given there are billions of them on earth today, but such figures refer to pitiful hybrid flocks who carry selectively bred, poor genetics for ever focused on providing a live-fast, die-young excess of cheap meat or eggs. Reared on intensive farms by truly massive agricultural conglomerates, the hens are kept out of sight and often out of mind. This industry is hand in glove with governments and supermarkets, and has led a deliberate and largely successful campaign to disconnect us from the joys of rearing our own livestock.

Confirmation of this sad fact is that it was an easier task to draw many of these different chicken breeds from admiring photographs of them than it was to seek them out in reality. But nonetheless I have tried to capture their movements and personalities, which vary hugely, as best I could.

There are fleeting moments, however, when the rare and pure breeds of chickens that remain can be seen by visiting poultry shows. Before the devastating twenty-first-century dawn of bird flu, a result of intensive farming, Britain in particular had a thriving interest in the showcasing of poultry breeds, a passion that hails back to the Victorian era and became known as 'the fancy'.

At poultry shows today, you'll find rows and rows of posing fowl and also their eggs on exhibition. The latter are beautifully presented on paper plates like perfect still lives, in a remarkable array of shell colours that nature intended eggs to be found in. Shades are as diverse as those laid by wild birds, in a range of pastel, white, cream, blue and many brown, toffee and chocolate hues, some even displaying a splatter of speckles. Some of the eggs are cracked open to inspect their colour and freshness. The winners are those with the freshest and most naturally golden yolk gleaming up from their egg white. Me and my grandma Sheila like to watch a TV show called *Four in a Bed*, which features bed-and-breakfast owners staying at one another's establishments and then commenting on their stay, creating huge drama. It amazes me to see the almost white scrambled eggs that are sometimes served at breakfast, often to great acclaim; I dread to think where the hens who laid such eggs live, as clearly it's somewhere without any grass!

The din of crowing from the cockerels at the shows is incredible, like dozens of alarm clocks going off in unison. To the chicken newcomer, a show can be overwhelming; to the chicken addict it is like being a fashionista viewing the windows in Bond Street!

And no bird is here by chance. Each one has been booked in by its fancier's breed club via the club's hard-working secretary, months in advance. Secretaries are vital for individual breed clubs, rallying breeders together with membership and meetings. Shows enable a coming together of people as well as poultry. Showing is a social time for many fanciers, a purposeful chance to show off their work of hatching and rearing, as their passion is an otherwise lonely pursuit with just the chatter of clucking for company. With the number of shows in decline, social media is today the saving grace for many pure and rare breeds of poultry, allowing fanciers to be in touch with one another and to swap eggs and stock.

Away from poultry shows, open farmyards and city farms are as important for old breeds of farm animals as good zoos are for endangered wild animals. The farmyard at Chatsworth in Derbyshire, a place I have known all my life, is one of these precious places. The chicken runs here hold many beautiful breeds on a grassy and wood-chipped slope. Each summer, broods of chicks are hatched monthly. They jostle about in their various groups, split by growth rates like classes of children at primary school.

The stable pens, with the written hatch dates on their doors, go from flocks of day-old bundles of innocent Easter-card chicks to gawky teenage rabbles, who, half grown and with their baby fluff fading out to their first feathers, have personas more like vultures. These older chicks are known as growers.

Through the wire doors, you can watch the little birds pecking about under their overhead heat lamps, each with the first promising gleams of feathers that are distinctive to its particular breed. I love the noise of cheeping and the sweet smell of wood shavings.

It is my hope that you will approach this book like a poultry show or an open farm. I have given my artistic and personal impressions of just a handful of the chicken breeds in the world, but I hope it will help you discover, or reconfirm, which are your favourites and encourage you to keep them yourself, or at least to remember them.

One last flight of fancy and an ode to the age-old question of 'What came first, the chicken or the egg?' In the case of rare- and pure-breed poultry, it is the fowl-mad fools who thankfully fall in love with, keep and continue to hatch them!

ANCONA

TYPE: Large fowl and bantam • EGGS: White
CHARACTER: Active, alert, hardy

Classy layers of white eggs in a beautiful dress of white-tipped, glittering feathers against a shimmer of beetle-green black

ANCONAS WERE ORIGINALLY bred in Italy, and a breed club was founded in Britain in 1898. The hens are of a light but deep body and will arise from the nesting box often as these are hard-working layers of white eggs. Broodiness is rarely an issue, with individuals too busy going about their business to contemplate the twenty-one days of steady sitting required for a clutch of eggs to hatch. If allowed to free range, which they really are made for, then they will forage for much of their daily diet during the spring and summer months.

The Anconas' black feathers are dramatically set off by the flicked dots of white, known as tipping. The older they get, the more white tips their new feathers gain with each moulting of the old ones.

Anconas are now more easily found as bantams rather than large fowl, which is true of many breeds, being more suited to smaller spaces and costing a little less to feed.

ANDALUSIAN

TYPE: Large fowl and bantam • EGGS: White
CHARACTER: Active, alert

An elegant layer, with blue and grey feathers like expensive roof slates and strong legs ready as a springboard for take-off

LIKE ALL MEDITERRANEAN laying breeds, who lay generous numbers of white eggs, the Andalusian is able to catapult herself up from the floor with ease thanks to her light body and generous wings, her lovely big comb flopping furiously as she does. She announces her eggs with a loud and long clucking known as the 'egg song'. Such hens are best not kept in suburbia; noise pollution letters may be sent swiftly to the council.

Andalusians are not common because you have to hatch many eggs to get birds with the breed's iconic blue feathers; a lot have black or white plumage with grey splashes instead. Whatever their feathers, all of them lay very well. The breed's genetics require understanding, but it's a good breed to keep if you want a challenge and are able to have them as free-range hens.

APPENZELLER SPITZHAUBEN

TYPE: Large fowl • EGGS: White
CHARACTER: Active, alert

Gold Appenzeller Spitzhauben

The forgotten cast members of *One Hundred and One Dalmatians*, these regal hens wear forward-pointing tiaras

ELEGANT BIRDS OF Swiss origin, the breed's full name is Appenzeller Spitzhauben, with *spitzhauben* the word for 'pointed bonnet', part of the traditional dress worn by Swiss women. Their arrival in Britain was thanks to Pamela Jackson, one of the six famous Mitford sisters, who lived in Switzerland in the 1960s and kept this native breed, having fallen for their charms. Pamela returned to England in 1972, smuggling a clutch of eggs from her beloved Appenzellers, hidden in a chocolate box apparently. She had several times asked the British Agricultural Minister for proper permission to import the breed but got no reply to her letters; she then took matters into her own hands. The precious eggs were hatched by her younger sister Deborah, Duchess of Devonshire, in the incubators at Chatsworth House, and these birds formed the foundation of the British stock of Appenzellers, the national breed club forming in 1982. Pamela was an example to all fanciers in being generous with her spare birds, enabling the breed to quickly gain many devotees.

The Appenzeller hen is light-bodied, always alert and very active, and lays white eggs in good numbers that have a rich taste due to the bird's constant pecking of green shoots and other naturally found morsels. Its comb is called twin-horned, appearing as red spikes through the forward-pointing crest feathers.

Appenzellers can take some time to settle into a place and some people remark that they either flourish superbly or simply don't thrive. In particular, they are intolerant of confinement. Free ranging is essential for their happiness, or a very large hen run, such as a fruit cage with its netting enforced with anti-fox galvanized wire – the sort of accommodation needed for all such flighty breeds that like to get into a good strut.

The best advice for all new hens is to shut them into their hen house for the first day and only let them out the following morning. This encourages them to return at dusk, without the stressful need to chase them around. But flighty breeds like the Appenzeller may well choose their own less secure roosting places, at least to begin with. If this happens, try to catch the birds to save them from the opportunist Mr Fox. Hens see poorly in the dark, so they can be caught more easily at dusk and you can then return them to the safety of the hen house. If you must catch fast-running birds in the daytime – best avoided – then a fishing net on a long pole is a good aid.

Silver Appenzeller Spitzhauben

ARAUCANA

TYPE: Large fowl and bantam • EGGS: Blue
CHARACTER: Active, alert, hardy, occasionally broody

**The famous original layers of blue eggs,
all the way from South America**

WHEN EUROPEANS FIRST gazed at the Araucana's blue eggs, they presumed that they must be duck eggs. The breed was brought to England by plant hunter Clarence Elliott, returning from his expedition to Patagonia in 1930 (unfortunately the single cockerel ended up in the ship's kitchen on the voyage back in a fatal miscommunication).

The breed's dominant blue-egg genetics have been used to produce blue- or green-tinted eggs in other breeds, including Cream Legbar and the popular Easter Egger in the US. Of the many types of feather colours, the Lavender, bred in the 1930s, is a favourite and their moulted feathers look especially chic next to the blue eggs.

The hens are good layers both as large fowl and bantams. They make very determined but, in my experience, aggressively hand-pecking broody hens. The eggs need to be collected often to help curb such keen sitters. Handle the chicks each day to help gain their trust and tame them, if not they'll often prove to be skittish adult birds, but they will thrive if allowed to range freely.

BARNEVELDER

TYPE: Large fowl and bantam • EGGS: Light brown
CHARACTER: Broody, confident, docile, hardy, robust

A robust yet subtly glamorous layer of caramel-brown eggs

COCKERELS USUALLY HAVE the more striking feathers, but with Barnevelders it is the hens who are arguably more beautiful than their male consorts. Each of her feathers is double-laced with a black outline, and her neck feathers are of a deep iridescent green that gleams in the sunshine. All this is complemented by the kind raspberry face of this very friendly breed.

Barnevelders were once hailed as one of the best layers of brown eggs. The showing of Barnevelders has focused more on their double-laced feathers and this has dimmed their merit of productiveness, but most hens will still lay a good 180 light toffee-coloured eggs in their first season.

As growing chicks, Barnevelders can be slower to sprout feathers than other breeds. They tend to go through a fuzzy fluff- and pin-feather stage for a few weeks before their first set of feathers frills out.

BELGIAN BANTAM

TYPE: True bantam • EGGS: White
CHARACTER: Broody, noisy, very tame, vulnerable to cold

Quail Barbu d'Anvers

Wonderful, playful little characters suitable for both farmyard and garden, in an excitable circus of mesmerizing feathers and body shapes

I WOULD LOVE Quentin Blake to draw a large family of these dear little funny bantams. Many keepers are charmed into such a flock by their huge and remarkable characters. Their small size means they are ideal for gardens, where they'll happily reside in large hutches and poultry arks that are too small for larger fowl, making them suitable for young keepers as well as adults.

This family of similar but individual Belgian Bantam breeds, named after the places they were bred, includes the Barbu d'Anvers, 'bearded from Antwerp', who dates back to the eighteenth century. Then there is the pert Barbu de Watermael, who has a little crest, the Barbu de Grubbe, who lacks any tail feathers at all and is known as 'rumpless' (rumpless breeds lack their parson's noses because they don't have the last two back vertebrae), and the Barbu d'Everberg, another rumpless breed, which also lacks wattles.

But it is the red-and-golden Barbu d'Uccle, from the outskirts of Brussels, who most has my heart, a wonder of an upright pigeon-sized bantam of Millefleur feathers. This 'thousand flowers' plumage is a delightful kaleidoscope of oranges, browns and reds splattered with a tipping of emerald green and blacks, and then set off by a sprinkling of white. The finishing touch is their sweet little feathered cheeks and their bibs, which are the 'beards'. These two features differentiate them from the otherwise identical Booted Sablepoots. Both of these breeds have outward-facing foot feathers that often require clipping back so the hens can about walk with more ease.

The cockerels of the d'Anvers and de Watermael need to

be handled often to stop them getting into any sparring habits, and these little birds fly well too. Broodiness will occur in all of the hens, who may lay about eighty small white eggs in their first year of coming into lay, but they aren't nearly as stubborn when it comes to wanting to sit them as the more popular Pekin Bantams.

All bantams are at great risk of rats, who will readily kill them, especially at night. Much can be done to avoid rodents, especially storing and using poultry feed with care. Invest in a secure feed bin and rat-proof poultry feeders. Have all poultry housing well off the floor and be tidy. Keeping the grass mown short is especially important as rats use it for cover. All this will help hugely in avoiding the perils of Samuel Whiskers.

Red Millefleur Barbu d'uccle

Lemon Millefleur Sablepoot

Porcelain Barbu d'uccle

BRAHMA

TYPE: Large fowl and bantam • EGGS: Light brown
CHARACTER: Broody, docile, friendly, quiet

The largest of chickens, with a steady,
quiet presence and dressed to impress
in their feather-legged pyjamas

QUEEN VICTORIA'S PASSION for poultry was so great that a grand aviary and fowl house was built in the park at Windsor. Completed by 1844, it housed the growing number of newly bred chickens that were regularly presented to her; it even included a sitting room so that the queen could sit at leisure with her avian subjects.

Among these breeds were the Brahmas. The first nine to come to Britain were sent to Her Majesty in 1852, courtesy of the American poultryman George Burnham, and delivered in a purple-and-gold box addressed to 'The Queen of Great Britain'. No doubt he was concerned that the cargo of his precious live birds could be neglected or lost; this paintwork did the trick. Their arrival was heralded in the *Illustrated London News* and they became the sensational celebrity chickens of their day. (It's easy to forget how such exotic new breeds might have caused a stir back then, now that internet search engines make every curiosity so accessible.)

Named after the Asiatic Brahmaputra River, Brahmas are giants among fowl. You wouldn't want a chicken any larger; the cockerels can often stand taller than most three-year-old children and need dog-kennel-sized accommodation rather than the usual hen house. They maintain a devoted following, but my heart is firmly with the Cochins. The two breeds are often confused by the untrained eye but are quite different. Brahmas in general have a more gamey look to them, in contrast to the more soft-faced Cochins. They aren't as fluffy in their trousers and their plumage is held tighter and closer to the body. You can tell that their plumage is more robust as the hens' looks suffer less from their cockerels' mounting passions! Brahmas also have a small pea comb not a single one.

There are many colours of the Brahma. The Light, otherwise known as the Light Columbian, resembles the classic black-and-white of the Light Sussex, albeit larger. The Buff Columbian is a rich honey colour and does look lovely, as does the classic Dark, which is like the colour known as Silver Partridge, a wonderful tapestry of greys with black lacing.

These relatively quiet birds are gentle Great Danes of the hen world and it is wonderful to see them dash about in their heavy-trousered foot plumage if they spy a handful of corn chucked down as a treat. The hens go broody and lay light brown eggs that are not as large as you might expect. The chicks mature slowly and grow well, though to reach their growing potential, they, like all breeds, need the best quality feed and plenty of access to short mown grass to peck at and a little sunshine for sunbathing. Opting for a cheap ration or expecting anything good to come from birds fed purely on table scraps (something that is, in any case, now considered illegal in the western world) will result in hens that are of poor condition.

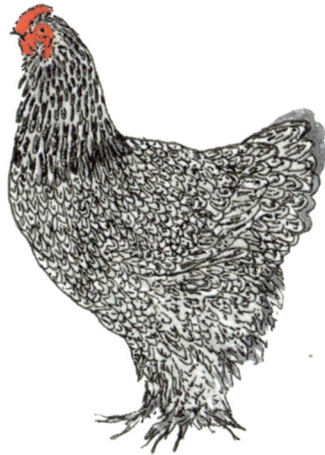

Dark Brahma

BURFORD BROWN

TYPE: Large fowl • EGGS: Dark brown
CHARACTER: Confident, hardy, robust

A classy and efficient layer of brown eggs with
impressively large yolks, now reared commercially

NAMED AFTER THEIR Burford homeland in the Cotswolds of England, these alluring hens are a household name thanks to their eggs being sold in supermarkets up and down the country. But one wonders how many people know what the hens actually look like!

Mr Philip Lee-Woolf is the exclusive breeder of the only genuine Burford Browns in Britain. These are descended from his grandmother Mabel Pearman's original stock of hens, which must have some Welsummer ancestry behind their much-admired eggshell colour. They lay notably large eggs with conker-brown shells, and in generous numbers for several years.

The hens have varying degrees of burnt-cinnamon red neck feathers that contrast against a striking black body of smart feathers. Friendly and calm, they relish being allowed to free range. They are my favourite hens to keep if eggs are the chief concern, and His Majesty King Charles III keeps a flock of them at Highgrove.

COCHIN

TYPE: Large fowl • EGGS: Light brown
CHARACTER: Broody, docile, friendly, quiet

Gold Partridge Cochin

Gentle, large hens with soft plumes, like an indulgence of moving pillows

COCHINS ARE VERY kind-faced and have a gentle presence, despite their huge size. The hen's soft body displays beautiful bouffant tail feathers, known as a cushion, and a huge amount of fluff. They'll do little harm within a mature garden of shrubs and roses because their leg feathers mean they can't scratch vigorously, though they will peck at plants. Their feathered pyjamas quickly soak up the wet and, alas, they don't seem to understand the concept of walking around puddles, rather than carelessly through them, which is especially bad news in cold winters. However, Cochins love to free range and will be happier for it than being mollycoddled. Give them a sheltered run with an attached hen house that is of a low bungalow design with a gently sloping ladder (they may be too big to go up steep hen-house ladders). A dry stable would suit a flock of Cochins very well for the winter.

Beware of the damage the large Cochin cockerels can do when mating the hens. Their claws and clumsy dismounting cause injury to the hens' softly plumed underwings. To prevent having half-plucked hens, Cochin fanciers are wise to keep their vigorous cockerels away from the hens once they have fertile eggs, to give the precious hens a well-earned rest. The same advice goes for the marital arrangements of other big breeds with soft feathers, especially the Orpingtons.

The eggs are pitifully small when compared to the hen's great size and not especially plentiful – expect little more than a hundred from a hen in her first season and fewer thereafter. Cochin hens do go broody and make attentive mothers, but the newly hatched chicks are at risk of being accidentally trodden on as the hens are quite clumsy with their big feet. Cochin chicks are hugely fluffy, like their parents, and they need to be given plenty of warmth for several weeks because they grow slowly. The feathers on their legs soon appear, looking like Elvis Presley's Vegas-show trousers, but I don't imagine these keep the young birds warm. If hatching Cochin eggs under a broody foster hen, choose one who is trusted to look after them for a while, and will not get bored of her chicks in a hurry. Keep the youngsters on growers' pellets until they are around twenty-five weeks old so that they grow well.

The Buff Cochins are the most iconic of the Cochin colours, thanks to them being reared by Deborah, Duchess of Devonshire, at Chatsworth House in Derbyshire. Peter Haywood, a renowned Cochin club fancier and the breed club's late president, supplied Chatsworth with the original birds, some of the best ever to have been bred. Derbyshire is a fierce place in the winter and spring, but the Cochins flourished here under the watch of the Duchess's poultryman Alan Shimwell. There are legendary tales of these calm birds being used as live décor during dinner parties: a fabulous change from the usual flower arrangements!

The Cuckoo, Blue, Partridge, White and Black are just as beautiful as the Buff and need more devotees; I personally have a great affection for them all.

Buff and Cuckoo Cochin

COMMERCIAL LAYERS

TYPE: Large fowl • EGGS: Brown
CHARACTER: Confident, friendly (when rehomed)

Rescued Ex-Commercial layer

Many a lucky ex-battery bird, the most domestic and hard-working of hens, today finds a happy retirement in allotments and gardens

DESCENDED FROM THE Rhode Island Red, these brown hybrid hens are the mainstay of commercial egg production. They are bred to be selfless layers, coming into lay as fresh eighteen-week-old pullets and understandably exhausted from this relentless task by the time they reach eighteen months of age, when most hens are sent to slaughter.

The abuse that these commercial hens face has been well documented but, shockingly, cruelty is not confined to just battery-cage farms. In the UK, nine hens per metre is the standard density for free-range production whilst so-called enriched battery cages can house up to 90 hens with each hen then having around 750cm^2 of space. The problem is not only lack of space. The most harrowing egg-farm footage that I've seen is an undercover investigative film that shows violent scenes of night-time chicken-catching gangs rounding up flocks of some thirty thousand-end-of-lay hens within the warehouse units of so-called 'free range' farms to take them to the slaughterhouse. Angry, deranged shouting could be heard from the gangs of men as they slammed and crushed armfuls of crying hens brutally into crates, throwing them around and at times kicking and even stamping on the pitiful, frantically moving carpets of these doomed innocents.

Some lucky hens, however, start new lives with professional hen-rehoming organizations who seek to engage with egg farmers. Likewise a number of independent family-owned farms have started to use social media to rehome their hens once they are no longer commercially viable, with great success.

The rehoming of these hard-working girls is a rewarding and worthy thing to consider but should not be done on a whim. The hens are usually badly pecked and will be afraid of being outside to begin with. But they will soon regrow their lost feathers, especially with the help of a rich layer's mash mixed with cod liver oil and baked beans (for protein) and they'll relish their retirements. However, whilst not ill, these hens are quite 'spent' as the industry terms them. Their bodies are weak from producing a huge number of eggs, and this often means that retirement can be short-lived. Some hens rehoused from poorly managed free-range systems may have got into the bad habit of feather pecking. The hens will need to be distracted and discouraged from doing this with anti-peck sprays that taste bitter.

I do have some conflicting feelings about the rehoming of ex-battery hens because the one downside of this worthwhile trend is that they take the place of the pure and rare-breed hens that are in great need of conservation. However, such birds make a great introduction to hen-keeping and no doubt result in many people going on to keep other breeds.

CORNISH GAME

TYPE: Large fowl and bantam • EGGS: Light brown
CHARACTER: Docile, liable to get fat, quiet

Curious, almost mythical fowl, these beady-eyed
characters move slowly on their sturdy
and stubby yellow legs

CORNISH GAME HAVE grumpily aged faces even as little chicks, trundling about like animated gargoyles wearing dark oil-green and black Macintoshes. This is a very friendly breed, happy to have their feathers stroked; they especially like to have their neck feathers tickled.

The breed's often laid-back attitude to life meant they were not used for fighting but instead are known for the quality of their meat, especially when crossbred with the plumper Sussex and Dorking. They were reared in Cornwall for many years, having come here from India due to the flourishing trade in tin between these places (first called Indian Game, their name changed to Cornish Game in the early twentieth century).

This breed must be allowed as much space as possible for ranging, despite being hardly able to get off the ground to perch up at night, in order to keep them fit and help avoid heart attacks. The hens lay at most about a hundred small eggs a year.

CROAD LANGSHAN

TYPE: Large fowl • EGGS: Brown with plummy sheen
CHARACTER: Broody, docile, friendly, hardy, quiet

Roly-poly hens with dark coal-house smutty
plumage and wondrous plum-shaded eggs

THESE IMPRESSIVE JET-BLACK plumed birds have a kind, docile nature. The cockerel's glamorous tail feathers are some of the largest to be found within the costume of mature cockerels, and the hens are fair layers of lovely eggs. They eagerly go broody and are quite careful sitters, considering their large size.

Major Croad brought these birds back to England from their native China in 1872, arriving on the heels of other large Asian breeds, the Cochins and the Brahmas. There was then a battle royale among the poultry press and fanciers about whether the Langshans were just Black Cochins. It would take thirty years and the efforts of Major Croad's daughter, with several imports of stock from China, to define that the Langshan was a distinct breed. It has a much shorter back, is less fluffy, with a firm outline to its body; it is far less profuse in its feathered legs when compared to Black Cochins. These days, it's a wonderful sight to see the pure-breed flock of Croad Langshans at Doddington Hall in Lincolnshire.

DEATHLAYER

TYPE: Large fowl • EGGS: White
CHARACTER: Active, alert, energetic forager, hardy

A rare German breed that has an excellent laying pedigree and a robust nature

THE BREED, THOUGHT to be some four centuries old, was first called Dauerleger, meaning 'day layer'. Over the centuries this name morphed into Totleger, which later became the more gothic-sounding Deathlayer. Whilst not, in fact, laying themselves into exhaustion and death, the hens are impressively productive, with around 200 white eggs in their first year. All hens lay fewer eggs with each passing year of their lives, but the Deathlayers and other European breeds especially can be counted on to lay decently and will earn their keep in eggs for several years.

This is a strong and hardy breed with very beautiful plumage, similar to the Gold Pencilled Hamburgs, and also comes in a silver colour. Greenfire Farms, a rare-breed hatchery in the United States that holds many rare European breeds seldom found elsewhere, imported Deathlayers from Germany in 2018.

This active breed does best at liberty, being constantly spirited and on the go. The hens can usually be dissuaded from broodiness, if needed.

DERBYSHIRE REDCAP

TYPE: Large fowl and bantam • EGGS: Cream, light brown
CHARACTER: Alert, energetic forager, hardy

Flighty characters that skip and run to take flight,
these good layers are an old breed from
the Dales of Derbyshire

RECORDED AS EARLY as 1848, the Derbyshire Redcap cockerel gives his breed its name thanks to his impressive comb, a marvellous, many-spiked fleshy cap that can grow to three inches long and looks like a red dahlia flower. Each plume on the back and breast of the hen is tipped with a half-moon of black spangling.

Hardy and independent, Redcaps really must be allowed to forage and range about the place to be seen at their best. Given their liberty, the hens lay superbly and their eggs are renowned for their very good flavour. They will seek out their own nests if their nesting boxes are not kept clean and dark enough. Well-thought-out, ready-made hen houses should have this in their design, but hens will soon tell you if they are not satisfied, scratching out the nesting straw and laying elsewhere. Nesting boxes in a converted shed or stable should be put in the darkest place, usually under the window, and some torn strips of old dark fabric or tea towels pinned across their entrance to act as curtains. A common mistake of novice keepers is to provide too many nesting boxes: one for every three hens is ample: hens don't require the luxury of having one nesting box each.

Mr W. H. Baker, a fancier of the Derbyshire Redcaps in the 1920s, said that half the breeders in the world lived in a ten-mile radius within the county. Despite many years of walking in the Derbyshire Dales I am, alas, yet to come across a flock of the breed pecking about a farmyard or foraging through the bottom of a roadside hedge; but I live in hope that one day that I shall.

DORKING

TYPE: Large fowl • EGGS: White
CHARACTER: Broody, hardy, quiet, robust

Red Dorking

Hailed by the Roman writer Columella as a bird of generous flesh, these hearty, hardy birds have white feet with five, rather than the usual four, toes

DORKINGS DATE BACK to the Romans. The market town of Dorking in Surrey, famed in the seventeenth century for hosting the greatest poultry market in England, was once this breed's stronghold; it was a more respectful time in the history of consuming poultry, when eating them was considered to be a luxury, rather than the tasteless commodity that chicken has become today. Dorkings are now a rare breed, even though they are still considered to have the finest flavoured and textured meat of any chicken. Not so long ago, tough old hens would have been used for casseroles and a wealth of stocks whilst spare young cockerels of breeds such as Dorkings reached a decent size for roasting or to provide generous leg meat.

Unlike modern table birds, which are ready for the table in under seven weeks, Dorkings take their time in maturing; it is an active breed that enjoys foraging. If roast chicken is the agenda, then the growing birds should be given twenty to twenty-five weeks of life on grassy pasture.

You can tell this is a breed of chicken with the potential to fit into a roasting tin when you observe this heavy bird's almost rectangular body, long back and short legs. The Dorking's feet have the breed's trademark of five toes, a dominant trait found only in four other breeds. The Silver Grey is the most well-known colour, but I would argue that the Red Dorking, with its plumes of glossy mahogany-brown, is more alluring.

The hens lay their white eggs well, in seasons of good light and weather, and the breed is docile. These stout girls will go broody and would once have been a popular broody hen to be trusted as foster mothers for the eggs of pheasants.

Dorkings were much used in poultry breeding to produce traditional dual-purpose breeds, including the Sussex and the Orpington. These breeds are both productive layers yet they still have a generous carcass at the end of their economical backyard lives, usually at four years of age, when laying eggs becomes a sporadic affair for old hens.

I rarely, if ever, eat chicken, but if I did then I would rear Dorkings. It may sound strange to preserve a breed by rearing the birds purposely for the cooking pot, but as the late Clarissa Dickson Wright said: 'If you want to save rare breeds, eat them, for goodness' sake, eat them!' Being a choosey carnivore – eating less but better, locally produced meat – can support small family farms who rear rare and traditional breeds.

Silver Grey Dorking

DUTCH BANTAM

TYPE: True bantam • EGGS: White
CHARACTER: Confident, easily tamed, surprisingly loud

Gold Partridge, Cuckoo and Yellow Partridge Dutch Bantams

Naughty little Punch and Judy characters with wonderful tail feathers spread out like fans

THESE ARE AS tiny as one would wish a hen to be and useful for the gardener, being a danger only to seedlings and good for insect control. A trio of a cock and two hens is a smart sight, their little breasts carried upright and proud – though the cockerel's crow is loudly shrill, despite his size. They fly surprisingly well and get up into the eaves of barns to roost.

The hens can use medium-sized terracotta plant pots on their sides to lay marble-like white eggs that hatch into chicks the size of large bumble bees. Keep the small family in the safety of an ark or rabbit hutch for several weeks, away from rooks, hawks and rodents. Drowning is also a risk. Provide water in shallow saucers to begin with, filled with grit so that the water is less than half a centimetre deep, or encourage the chicks to use nipple drinkers. Hand-reared chicks become hens as tame as budgies. I knew one plucky hen who laid her tiny eggs in her keeper's desk drawer. She would peck at the door, wait to be let into the office and then demand the drawer to be opened with spirited clucking!

FAVEROLLES

TYPE: Large fowl and bantam • EGGS: Cream
CHARACTER: Broody, docile, quiet

Grand, gentle dames among hens with a cappuccino froth of warm brown and cream feathers

A WONDERFUL, DOCILE garden hen, the Faverolles is named after the village in France from which the breed hails. They were originally bred both as table birds and as good layers, and it's thought there are both Dorkings and Houdans in the breed's ancestry as they are broad hens with a smoothly rounded body shape, and have light foot feathers that conceal the breed's five toes. The Salmon, as drawn here, is the most popular colour.

The breed's most striking feature is a wonderful bobbing beard, which is classed as being 'fully muffed', a term for when little feathers extend up and past the bird's earlobes and beneath the beak. The chicks hatch with these adorable little powder-puff cheeks. You may need to give the beard the occasional warm flannel wash and then a delicate drying using a hairdryer during the winter months if the wet and dirt of the ground sticks badly to these feathers.

Every year Chatsworth House, the palace of the Derbyshire Peak District, is decorated for Christmas – an idea of Andrew Cavendish, the 11th Duke of Devonshire. One memorable year the theme was the carol 'The Twelve Days of Christmas'. As part of the festivities, a hen ark was surrounded by an adornment of glistening lights and baubles, and here sat three Salmon Faverolles hens from Chatworth's farmyard, as the 'three French hens', the gift given on the third day of Christmas. They looked perfectly at home clucking softly along with the merriment.

FAYOUMI

TYPE: Large fowl • EGGS: White
CHARACTER: Alert, hardy, wild

Ancient and almost feral, this streetwise,
independent chicken carries precious genetics

A BIRD OF ancient Egypt, the Fayoumi survives in the chaos of the modern world, scratching out a largely independent living, coping with heat and meagre rations, with its strong genetics making it resistant to many diseases. They have long been kept in the Fayum district of Egypt for their egg production and were brought to Britain in 1984 by Victoria and Michael Roberts for their esteemed collection at the Domestic Fowl Trust.

These small, thin birds 'walk like an Egyptian', as the Bangles song goes, forward-thrusting and with a jiggling movement of their necks. They can run or fly off incredibly fast to get themselves out of harm's way. If caught, a Fayoumi cries with the din of a guinea fowl, and the male chicks sometimes crow at just six weeks. The hens like to lay in secret places and are mindful of their position in the pecking order, bickering often.

FRIESIAN

TYPE: Large fowl and bantam • EGGS: White
CHARACTER: Active, alert, energetic forager, hardy

Elegantly poised, with unique vanilla-honeycomb feathers and dashing red combs, this composed little hen lays her white eggs in very good numbers

PRODUCING SOME 240 eggs a year, Friesians are remarkably good layers, earning them a nickname in their native Holland of the 'Dutch Everyday Layer'. They get along with life happily without complaint, and don't seem to often go broody, quickly going off the idea if their eggs are readily collected.

This light-bodied little hen, adept at flying over walls and hedges, will prove to be very independent and forage far and wide. Grit should be available to all hens, for egg strength and digestion, and hard-laying breeds like the Friesian will eat a surprising amount. The best grits are those mixed for racing pigeons, made up of not only oyster shell but also silex stones, charcoal and red stone.

As for colours, Chamois Pencilled is the popular Chanel suit of choice for any discerning hen. The cockerel's white tail feathers look especially striking, carried proudly and arching off from his rich treacle pudding-coloured body.

GOLDTOP

TYPE: Large fowl and bantam • EGGS: Cream, light brown
CHARACTER: Docile, friendly, very often broody

Nature's incubator: the best and most trusted of willing egg-sitters

THESE ARE HENS of buff and sandy plumes, with a light sprinkling of black neck feathers, and burgundy wattles and combs, often with a sweet little head crest. From their looks, it's possible that a Goldtop was Beatrix Potter's inspiration for her character Sally Henny Penny in the tale *Ginger and Pickles*, illustrated among the rabble of curious customers within the chaotically run shop.

The breed is the result of crossing a Golden Silkie cockerel with a Light Sussex hen. They will sit all manner of eggs – before incubators, Goldtops and similar bantam crossbred hens were much valued by gamekeepers for the hatching of pheasant and partridge eggs. Their feathers are to be more trusted than those of the soft, fur-like pure-bred Silkies that can sometimes wrap around new chicks with fatal consequences.

All broody hens, especially those as dedicated as the Goldtop, must be encouraged to take a daily break from the nest to feed and dust-bathe if they are to be in good health when the chicks hatch.

GRONINGER MEEUWEN

TYPE: Large fowl and bantam • EGGS: White
CHARACTER: Active, alert, hardy

Beautiful tigress-striped hens that are excellent layers
for either a large hen run or for free-ranging

THIS DUTCH BREED is a hard-working and hardy layer of white eggs in both its large fowl and bantam miniature forms (but be aware that the bantams tend to be more prone to stubborn broodiness than the large fowl version of the breed). The breed nearly became extinct in 1980, with numbers falling as low as twenty-six birds. Luckily some fanciers in the Netherlands saved these chickens and today they have a breed club in their homeland that maintains a good standard for both the bird's looks and productivity.

The Groninger Meeuwen is closely related to another Dutch breed, the Friesian, but is a little larger in the body frame, with a slight slope to the back, and the bird doesn't hold its tail feathers in the same upright manner. The black pencilled markings of both the hens and the cocks are set against the breed's three different base feather colours: Gold, which is more of a deep blood-orange shade, Lemon and Silver. The hens have bonny red faces and a traditional single comb.

HAMBURG

TYPE: Large fowl and bantam • EGGS: White
CHARACTER: Active, alert, energetic forager

An independent, good-laying hen that's fit for a catwalk, sporting a dress of spangled dots and a raspberry comb

THESE OLD WORLD, elegant chickens, once known as 'everlasting layers' or 'endless eggers', are still productive and deserve to be kept much more than they are at present. They have no connection with Germany, despite their name, but are thought to originate in the Netherlands and to have spread with the Vikings, possibly passing through the port of Hamburg before reaching Britain – the exact story seems to be a mystery. Their feather colours were then refined in Lancashire and Yorkshire during the nineteenth century in the boom time of the fancy of poultry. This is a beautiful breed in all their colours, which come in Silver Spangled, Gold Pencilled, Gold Spangled and also Black, which was once common and known as the Black Pheasant Fowl.

Tough cookies, with Old English Game blood in them, they certainly have some good brain cells. If you don't pay much attention to them as chicks, a flock will be flighty and spirited. They look absolutely stunning if given the free-range liberty of a yard, busily running about and clucking, and they will delight in rustling about on a kitchen garden's compost heap.

They aren't stubborn as broody hens and you can discourage them from sitting by regularly disturbing their hen house as you collect eggs and clean it out. But this breed has good survival instincts, so expect some brooding in the summer as the hens know their survival depends on at least one of their flock hatching a clutch of eggs. The rest of the year, the hens lay their white eggs remarkably well and any chicks that are hatched mature quickly and soon learn to use their little wings to perch up safely at night.

HEDEMORA

TYPE: Large fowl • EGGS: White
CHARACTER: Energetic forager, hardy

An old and hardy breed of farmyard fowl, adapted
to Nordic cold, recently brought to Britain

THE HEDEMORA IS a small hen that lays light brown eggs in good numbers and has a docile and active nature. Native to Dalarna, a part of central Sweden where the landscape has been shaped for centuries by many small arable farms, this is what is known as a landrace breed: one that has evolved to suit its place with little interference. Such strong genetics make it very hardy.

The only birds to be found in the UK arrived thanks to the efforts of Hens on Oxney in Kent. There is no breed standard, but the birds' feathers do have a range of differing and recognized styles. Some of them resemble the fur-like feathers of a Silkie but in a coarser fashion, providing the best insulation nature could give against winter temperatures that often fall below freezing. Most of the Hedemora have some feathering to their feet too.

The trimming of a cockerel's spurs should be done seasonally, using a pair of good toenail clippers and a nail file. Unless they are very long, take no more than an inch off the end to ensure the sharp point is blunted. This won't hurt the bird, but it will prevent the spur growing up and in towards the leg, and also prevents the cockerels doing damage to the hens' backs when mating.

HOUDAN

TYPE: Large fowl • EGGS: White
CHARACTER: Docile, friendly, quiet

A fancy pantomime-costumed rare-breed hen
that is very tame and an excellent layer

THIS GENTLE HEN suits garden life very well. Bred in France from the eighteenth century, she has a similar crest to the much better-known Poland but is less flighty, being heavier in the body because she was bred for the table as well as for laying. The hens often lay their white eggs into the winter months despite the lack of day length, which is unusual; without artificial lighting, most hens reduce their laying to take a sensible and well-earned rest, laying again once springs returns.

Because of their crests, the Houdan is best kept with other docile birds to prevent any bullying, and you should carry out regular inspections of the crest feathers for any issues with mites. These are easy to spot on light feathers but harder on dark-feathered breeds such as the Houdan. I've found the best powders to deal with mites are those sold to deal with bed bug infestations!

JAPANESE BANTAM

TYPE: True bantam • EGGS: White (and tiny!)
CHARACTER: Chatty, friendly, vulnerable to cold and wet

Also known as Chabo Bantams, the cockerel's iconic tail feathers, arching like a wonderful sail, are as long as this little bantam's body

WITH THEIR TAIL feathers carried like *uchiwa* fans, Japanese Bantam hens have the poise and striking presence of geishas. They have a regal air, despite their size, and appear in Japanese artworks from the early seventeenth century.

You will have to be good at mowing the lawn if you keep Japanese Bantams because these truly ornamental chickens cannot cope with untidy, long grass due to their incredibly short legs. Rather than strutting around like other hens, they instead seem to float along the ground.

The hens lay tiny white eggs in small numbers and often go broody, making good mothers and being the perfect size to fuss over their tiny offspring. But alas their stature affects the fertility of eggs, as with the short-legged Scots Dumpy, and not all the chicks succeed in hatching.

With care and good, dry quarters, these little birds often live long lives and have strong personalities. The hens, especially, have a dear little look about them, always seeming a little concerned.

JUNGLE FOWL

TYPE: Wild ancestor of domestic chickens • EGGS: White
CHARACTER: Alert, very good fliers, wild

The wild ancestor of all chickens, often spotted
in the background of nature documentaries
but rarely given an official mention

TO SEE A jungle fowl take flight with ease makes you realize that chickens were once proper birds. It is incredible to think that all breeds of chickens descend from this little pheasant-like bird, domesticated more than four thousand years ago in the jungles of Asia, and that all chickens today still carry their ancestral instincts.

Jungle fowl like to lay eggs in secluded nests, to perch at night for safety, and to scratch and dust-bathe at will. The cockerels are truly dashing, with the fast movements of swashbucklers. The jungle fowl's woodland habitat means that all chickens to this day like the secure feeling of having branches around them and a choice of spending time in both sun and dappled shade that's dry underfoot. In a garden setting, hens adore shrubberies, fruit bushes (both of which are improved by the nitrogen-rich manure of the birds' droppings) and scratching about underneath hedges.

The jungle fowl is rare today, in terms of pure genetics, and is kept in zoological collections as part of captive breeding programmes for their precious DNA. They are rarely found in a domestic setting, being more suited to the aviary accommodation often reserved for ornamental pheasants. Out of all the chicken breeds, the Gold Dutch Bantams best resemble their ancestors; the cockerel's plumage especially so.

LEGBAR

TYPE: Large fowl • EGGS: Sky blue
CHARACTER: Active, alert

Superb layers of blue eggs, these pretty
hens have feather-tiara crests

THERE'S BEEN A rush of popularity for the Cream Legbar because the hens are prolific layers of blue eggs. The only complaint I have is the word 'cream' in the name. If true to their breed standards, they do not resemble this colour, apart from their breast feathers. Their correct feather coloration should be various shades of grey, but 'Grey Legbar' does not really sound as idyllic!

The breed was developed from the blue egg-laying Araucanas. Professor Reginald Punnet of Cambridge University, who was keen on poultry breeding and wanted to increase the number of useful breeds available to British farmers, created the original birds. In 1947, the first Cream Crested Legbars were shown at the London Dairy Show, originally held annually at the Olympia in London, where many newly created breeds of livestock were introduced to great crowds.

Cream Legbars can be quite runty in their size and it is best to hatch chicks from the largest eggs. The chicks are famous for being what is called 'sex-linked', which means it is easy to distinguish males from females as the eggs hatch. The vanilla-yellows mixed with some light grey fluff will be cocks and the buff-brown striped ones will be hens.

The breed thrives on having lots of space and can be aloof at first, easily flapping into trees to roost come dusk, but they soon settle down once they become familiar with their keeper and new surroundings.

If any of the hens do go broody, they can usually be encouraged to forget the idea – there might be some sharp pecking to your hands – but they make very good mothers if allowed to sit their eggs. They will prove to be outstandingly friendly hens if reared from chicks by either yourself or a tame broody hen.

Close cousins of this breed now have a large commercial presence. From these Cream Legbars, the more recent Old Cotswold Legbar was bred by Philip Lee-Woolf in the 1990s. The breeding programme took the commercial laying lane from the outset, and this strain lays a slightly larger egg of a uniform size. Showing the birds as a pure breed wasn't a priority, hence the birds' unstandardized feather colours, which vary in a wonderful array of butterscotch buffs and browns, often with impressive head feathers too.

Lee-Woolf's blue- and pastel-coloured eggs caused a sensation when they were first sold in Harrods and Fortnum and Mason, and are now stocked in supermarkets, with the breed trademarked to a commercial egg enterprise. Today, the majority of Cotswold Legbars laying for supermarkets are a white-feathered fowl due to them being crossed with White Leghorns. This cross is known as the 'Snowbar' and the dainty hens carry a white crest due to their Legbar ancestry.

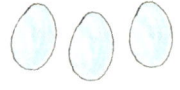

LEGHORN

TYPE: Large fowl and bantam • EGGS: White
CHARACTER: Constantly alert, energetic forager

Exchequer Leghorn

Prolific layers of white eggs, sometimes dizzy-headed but elegant when composed, with large floppy combs

NATIVE TO ITALY, the breed instantly found favour in America after coming over from Tuscany in 1828 because they lay large white eggs with great generosity. The Leghorn came to Britain in 1853 but didn't take off in the same way because the British traditionally prefer a brown egg, whereas the American market likes white eggs.

In backyard situations, the Leghorn is a very hard-working, pretty and ungreedy hen, scratching happily all day with her yellow legs and laying an egg almost every day. The hens go broody rarely, if ever. The Brown, Lavender and Exchequer are beautiful and, like the most common colour, the White, they carry big white earlobes like blobs of marshmallow on each side of their faces. The cockerels have great big single combs and wattles, some of the largest to be seen. Often the hens are mistaken for cockerels, from a distance, as their combs are very large too.

The Leghorn, unfortunately, was destined for the world of intensive farming in the United States because it was good at laying from the outset. When you hold a Leghorn hen, it is incredible to think that her slender body can produce a minimum of 250 large white eggs a season in natural conditions, and dozens more under artificial lighting. This is why the genetics of Leghorns are found in almost every battery-laying house throughout America, where slender hybrid birds are slotted barbarously into cages. Huge factories (they can no longer be considered farms) hold staggering numbers of hens. Cal-Maine Foods, the largest producer of fresh eggs in the US, was reported to have culled some 1.6 million hens in Texas alone in the spring of 2024, due to outbreaks of avian flu; this was just 3.6 per cent of the company's flock. Governments and the egg industry have blamed the backyard keeper and family farm for being a risk to commercial birds, regarding this virus, but the pattern of outbreaks mostly occurs on huge factory farms, although such facilities are championed by industry as being bio-secure. Most harrowing is the method of mass culling for these warehouses full of hens in the event of an outbreak. It is known, chillingly, as 'ventilation shut-down'.

Thankfully not all commercial Leghorn hybrids face such brutality. In Stow-on-the-Wold, my friends Paddy and Steph of Cacklebean Eggs have developed a lovely Leghorn hybrid called the Arlington White. Happy hen principles are hard to find today, aside from independent, family farms such as Cacklebean, who are not shackled by industry and supermarket requirements. You can sometimes see their free-ranging flocks foraging from the road, a sea of little snow-white bodies with huge strawberry-red combs breaking up the green pasture. Every morning, the birds march out from their hen houses with a determined disposition, having diligently laid their white eggs. These are then hand-collected and boxed into Cacklebean's unique egg boxes that open upwards, like a box of luxurious chocolates.

Lavender and Exchequer Leghorns

LINCOLNSHIRE BUFF

TYPE: Large fowl • EGGS: Light brown
CHARACTER: Broody, calm, confident, hardy, robust

A hardy rare breed with golden-treacle feathers,
earning its keep as a plump layer of tinted eggs

VISIT THE FLAT terrain of the Lincolnshire countryside and a trained eye cannot help but notice the soulless vision of the area's many factory chicken farms. These compounds are made up of identical low, windowless sheds, grouped by the half-dozen, if not more. Inside each one are tens of thousands of pitiful broiler chicks living out their short lives to supply Britain with cheap and tasteless chicken. More than thirty-one million of these 'broiler' birds are reared annually in the county of Lincolnshire alone, according to Compassion in World Farming.

With such a local blight of factory farming, it is a refreshing and hopeful sight to see the flock of rare-breed native Lincolnshire Buffs strutting through the lush pastures of the South Ormsby Estate's walled garden. The Lincolnshire Buff is a rare breed today and this is its stronghold.

First bred in 1900 by crossing Buff Cochins with Red Dorkings and Old English Games, the Lincolnshire Buff was once popular as a table fowl, with generous legs and breasts. But the breed then almost vanished until local fanciers teamed up with Riseholme Agricultural College to revive it in the 1980s.

The breed has a longer body frame than the Buff Orpington, is less fluffy and more of a muddy colour, with a golden-treacle shade of buff in their feathers. The calm, charming, large hens lay a decent number of light brown eggs; they are an easy-going breed good for backyards and allotments. They are good at sitting and rearing large broods and their chicks grow quickly, the spare cockerels making good table birds.

MARANS

TYPE: Large fowl and bantam • EGGS: Brown, dark brown
CHARACTER: Broody, confident, hardy, robust

Neat and plump speckled hens, layers of admired brown eggs, with huge, rugby-player husbands sporting great combs and wattles

THESE BLOSSOMING HENS are famed for their brown and often speckled eggs. The classic Marans colour is Cuckoo, with each feather having an outline of very dark, almost black-grey, fading towards the middle, with an occasional sprinkling of white.

Then there are Copper Black Marans hens, who potter about in the darkest of chic funeral attire and have a few little tufts of feathers on their black legs. They lay what really does look like a chocolate egg at the start of each season, with the first few eggs always the darkest. The shell pigment fades to more of a hot-chocolate powder brown until the birds have a good rest or moult, and then these darkest of brown eggs return. The bonny, cappuccino froth-coloured Wheaten Marans lays an even richer cocoa-coloured egg with an orange clay-like hue.

The competition for the best of these eggs is fierce at poultry shows and it is a wonder to behold the showing tables burgeoning with such egg entries. As with all breeds, it's the cockerels' genetics that play a vital role in ensuring the best coloured eggs, so be sure to keep those that carry especially good eggshell colours in their pedigree.

Marans hens are all confident characters and very friendly towards their keepers, but they can be dominating towards more timid breeds. These aren't bouncy hens, and they will stay within the fenced confines of an allotment or garden but need space to ensure they exercise well and don't become fat – easily done if you feed too much corn to this French table breed. Expect occasional broodiness, but the hens are docile sitters and very good mothers if they are permitted to sit on a clutch of eggs.

The Marans cockerels are mighty, red-blooded males who ooze testosterone and will be quick to become quite bolshy if given the chance. But there is no excuse for allowing one to become aggressive. Regular and friendly handling will ensure he knows who is boss. A good, weekly underarm cuddle is the thing for all cockerels. It is imperative to never strike back when dealing with an unruly cockerel. Cuddling is what pacifies them, so handle young male birds often to prevent aggression becoming a problem.

As youngsters, it is easy to tell the cocks from the hens. The young boys have much bigger legs and noticeable large combs that are the soft pink of uncooked bacon, and their feathers are much more of a muddled chessboard pattern of black and white, whilst the little hens' plumes are often much darker, with very little white.

Wheaten Marans

MARSH DAISY

TYPE: Large fowl • EGGS: Light brown
CHARACTER: Hardy, robust, energetic forager

A hardy and calm rare-breed hen who
lays well and has a friendly nature

THIS MARVELLOUSLY NAMED rare breed has a flat cap of a rose comb and a small but pleasingly plump and robust body. The hens are calm-natured, quickly becoming friendly in their personalities, and are a good choice for urban hen houses. They lay light brown eggs of a rich flavour well into old age. The cockerels have a varied reputation. Some find them to be a little too keen on pecking boots, while others regard them as being docile in their manners.

Marsh Daisy hens are said to hail from Southport in the north-west of England, where they laid well despite the damp, compared to other breeds – although no chickens truly thrive in wet or muddy conditions. A Mr Wright first bred them in Marshside in Lancashire in the nineteenth century, crossing Leghorns with Hamburgs and Old English Game birds. Then a Mr Charles Wright from Doncaster further added to the mix by bringing in the blood of Sicilian Buttercups. By 1920, the breed had become known as a good breed both for laying and for the spare cockerels being worthy of rearing for the table. Alas, like many old breeds, this fame was short-lived and they were never massively popular outside their original area. The breed was brought back from almost completely disappearing in the 1970s.

The Marsh Daisy is a bit of a poster girl for the Rare Breeds Survival Trust organization; they are often seen pecking about the RBST's stall at agricultural events in an attempt to raise much needed enthusiasm for dwindling breeds. Surprisingly, given their rarity and narrow geographic range, there are some five different colours of Marsh Daisy, but the most easily found seems to be the Wheaten variety.

NEW HAMPSHIRE RED

TYPE: Large fowl and bantam • EGGS: Light brown
CHARACTER: Confident, hardy, robust

Close American cousin of the Rhode Island Red,
but in a lovely tomato-soup orange plumage

EASY GOING AND reliable, the New Hampshire Red exists thanks to the work of the University of New Hampshire. It was recognized as its own breed in 1935, rather than just being seen as an orange-coloured version of the Rhode Island Red: the birds are a little heavier in size and a little less hawkish-looking. They arrived into Britain in the 1930s and, like the Rhode Island Reds, were crossed with the British Light Sussex hens to produce commercial pullets. The hens are hardy and docile, and make for very good layers. A flock of both these breeds mixed with the Barred Plymouth Rocks would be a fine tribute to the pure-poultry breeding efforts of the United States.

Because the breed doesn't have leg feathers, which is typical of all the well-known laying breeds, it is conveniently easy to spot and treat the common poultry condition known as 'scaly leg'. The legs can be smeared over with Vaseline once a year to keep them looking their best by suffocating this microscopic parasite.

NORFOLK GREY

TYPE: Large fowl • EGGS: Light brown
CHARACTER: Energetic forager, hardy, robust

A rare breed from Norfolk, with a body
of black feathers complemented by a silver,
white-laced neckerchief

THE NORFOLK GREY was bred by Mr Fred Myhill of Norwich in 1920. Unfortunately for these chickens, he chose to name them Black Marias, after the notorious German artillery shells. This didn't sit well with many people and they were never widely kept. In 1925 the breed was renamed the Norfolk Grey, but this wasn't enough of a relaunch. They almost became extinct by 1960, but a trio was found in Banbury and formed the basis for the breed's revival. Numbers have recovered enough today to produce an active breed club page on Facebook.

This is a hardy breed, causing little fuss, and will be grateful to be kept by anyone who wants a good all-rounder backyard fowl. The quite heavy-looking hen has nice upright tail feathers and is a good layer of pale brown eggs.

OLD ENGLISH GAME

TYPE: Large fowl and bantam • EGGS: Cream
CHARACTER: Active, broody, confident, territorial

Courageous and constantly alert to rivals, with a long and often bloody history throughout Britain

COCK FIGHTING, A 'sport' since Roman times, was outlawed in England and Wales in 1849 under the Cruelty to Animals Act. Before this, a good stud of Old English Game was as precious as a racing horse, and the practice of rearing Old English Game birds ran deep. Once the ban became law, there was a huge shift towards breeding poultry for showing instead. Almost all British breeds have a bit of Old English Game in their genetics.

In old paintings, you see fighting birds prepared for sparring, with their wattles and combs 'dubbed', or cut off, to prevent bloodshed. Their beautiful, large tail feathers were also severely cut back, along with their saddle and primary flight feathers, so rival cocks had little to grip onto with their beaks as they fought. Once the showing of the Old English Game took over from cock fighting, the very finest of feathers mattered most, and they still do today. The breed's stiff feathers, in a huge array of colours, are also much sought after for fly fishing because they float well and look like a fly on the water.

Old English Game cocks fight at any opportunity but, despite this, prove to be amenable to their fanciers and behave well at shows. The handling of these birds and the relationship between fowl and fancier is an art form in itself; there are protocols for this aspect that are judged at shows.

Aside from feeding, the birds need little assistance once introduced to a decent assembly of hodgepodge outbuildings for shelter and roosting. They will often find their own safe roosts and inaccessible places to lay their eggs, well away from any hen house. The hens do a good job of sitting and rearing their chicks and they'll turn up very proudly with a day-old brood in tow. Be warned that an Old English Game hen with her chicks will often fiercely attack any curious hens who come near and readily take on any prowling rats. Fighting and bickering can be common among growing chicks too, so a large space is needed at all times, and they are best not reared with other breeds as they are dominant, even as infants.

There are many different strains of Old English Game to be found, and they have been much written about. There are at least thirty colours. I must admit that they are not at all my type of bird, though the strong genetics and independent spirit of the bantams have to be admired. They are punchy in their demeanours and can cope when circumstances result in them living an almost feral existence.

Another word of warning. Those keeping Old English Game should keep their best stock under lock and key. Just because cock fighting is banned doesn't mean it doesn't happen; the theft of unguarded Game birds is commonplace. One occasionally hears of people keeping cocks for illegal fighting being raided and awful scenes being uncovered of these great birds in small and shabby cages.

OLD ENGLISH PHEASANT FOWL

TYPE: Large fowl and bantam • EGGS: White
CHARACTER: Alert, energetic forager

An elegant, hardy and poised breed, suited to free-ranging, constantly scratching at the ground yet alert to any threat

BEAUTIFULLY SLIM AND light on their feet, but not as nervous as other light traditional breeds, the Old English Pheasant Fowl has been a fowl of the Yorkshire and Lancashire uplands since the seventeenth century.

In their day, this breed was a good all-round addition for the country farmyard. They are able to forage well and fly out of harm's way but aren't too wild, and the hens will go to the hen house to lay a generous number of their white- to cream-coloured eggs. They do go broody occasionally and will do a good job as mothers, but they'll often be nervous sitters. Spare cockerels have an excellent flavour as table birds, though young cockerels that are not welcomed by the dominant male may choose to roost in nearby trees or outbuildings – if they are lucky enough to find such places on offer as bachelor pads!

The breed's feathers have what is called horseshoe lacing, with breast feathers of a rich black-green set off by a mahogany-brown turning to a cinnamon-orange in the sunshine, and the cockerel's tail feathers are long and flowing. Their rose combs, particularly of the cocks, are a statement of sea urchin-like spikes.

ORPINGTON

TYPE: Large fowl and bantam • EGGS: Light brown
CHARACTER: Broody, confident, docile, friendly

Jubilee Orpington

Regal, plodding ladies in feathers fit for the stage
of a Broadway musical, this famous breed requires
larger than average nesting boxes

THE BUFF ORPINGTONS, pictured overleaf, are icons of pure-breed hens. There are chickens . . . and then there are Buff Orpingtons, with their golden and light saffron-yellow fluffy feathers carried off with a sophisticated oomph of elegance. Bred in 1894, they are now regarded by many as being the Golden Retriever of the poultry world, much beloved and with a calm and friendly disposition.

Orpingtons are quiet birds and make for very good, large broody hens. They have an advantage over Buff Cochins in having no feathers on their legs, making them adept at coping better in wet weather, but this is a disadvantage when you have Orpingtons at large in the garden because those bare legs will make short work of creating dust baths in your flower beds. The bantam miniature versions of the large fowl are a better choice for the more conscientious of gardeners and, being smaller, are reliable at organic weed-control between your rose bushes.

Queen Elizabeth, the Queen Mother was gifted a trio of Buff Orpingtons in 1977, the same year that she accepted the invitation to become patron of the British Poultry Club. One can imagine her devoted steward William Tallon ('Backstairs Billy') shooing them about the Clarence House parterre – along with the corgis – whilst holding trays of freshly prepared gin cocktails.

Before the Buff came the Orpington's original colour, the Black, reared by William Cook in Kent in 1886. Their iridescent feathers shimmer into a beetle-green velvet in the sunlight. Cook later presented his Jubilee Orpingtons to Queen Victoria to honour her Diamond Jubilee in 1897.

The Jubilee coloration is absolutely marvellous, with feathers the colour of a very rich fruit cake tipped with white. The Lavender Orpington is gaining popularity but few strains of this newly bred colour at present carry off the much-admired fluffiness of the more traditional Orpington colours, so you'll need to search to find the best of these birds.

The Orpington was once a very good layer and a provider of good white breast meat. Today they are kept more for looks and there are only a few productive strains remaining, rather than those that are just for the show tent.

In terms of housing, a bungalow-type design will be appreciated, and a converted summerhouse ideal. Be careful to not overfeed Orpingtons with corn; they need the range of a lawn each day so they do not become too fat and suffer from the fatal fatty-liver disease that is a problem with overindulged pet hens.

More generally, suppliers and hen keepers are struggling with compound feed because much of it contains soya and this can come from deforested or cleared land in South America or have other environmental issues. A recent addition to the poultry-feed market is the option of sustainable Black Soldier Fly larvae, sold live and marketed as ECOnourish, which is rich in protein. Pet hens become completely addicted to this beneficial chicken caviar, and it doesn't make them fat – if you can cope with all the wiggling!

Buff, Black and Lavender Orpingtons

PEKIN BANTAM

TYPE: True bantam • EGGS: Cream
CHARACTER: Broody, confident, docile, friendly

A favourite little hen with a strong and often mischievous personality

PEKIN BANTAMS CANNOT fail to bring a smile. Delightful little characters with huge personality and confidence, they bustle excitedly about the floor with their bouffant tail feathers in a neatly cupped beehive hairstyle. They positively adore being adored and look very happy at poultry shows, where the best examples wait to be judged like animated tea cosies. The best Pekin fanciers are those who are brave enough to allow their prize-winning charges enough time out on grass so that their faces colour up to be a lovely shade of raspberry jam. Show birds who are kept inside too much, for fear of getting their foot feathers damaged, always look comparatively pale and much sadder in their combs and wattles. Weekly outings on fresh grass are very beneficial for all the fancy breeds.

Pekin Bantam hens make ideal garden hens and are happiest as mothers, doing an attentive job of rearing chicks. The cockerels are usually real gentlemen to their hens and will tend to join in family life, waiting expectantly by the nesting boxes of their sitting wives.

Pekin hens are very persistent broodies, a trait that can come into conflict with their popularity as pet hens. This devotion – at times, a brattish determination – to sit eggs, is best described in D. H. Lawrence's *Lady Chatterley's Lover*: 'There is no self in a sitting hen.' New owners need to be prepared to cope with this and ensure that hens are not allowed to just sit on empty nests and become anaemic as they grumpily pine for eggs to hatch, putting them also at risk from mite attacks.

When they are not obsessed with the idea of hatching, the eggs you get from Pekins are small and cream-coloured and in fair numbers for a fancy bantam, with about a hundred in their first year of coming into lay.

Buff Pekins (one of the oldest of Pekin colours) originally arrived in Britain in 1860 as loot from the Imperial Palace Gardens of Peking towards the end of the Second Opium War. British and French soldiers scaled the 15-foot walls that enclosed its 'gardens of perfect brightness' and mercilessly plundered the contents. Pekingese dogs were also stolen and these share the Pekin Bantam's plucky character and mop-like appearance.

Today there are many colours of Pekin bantams. My personal favourites are the Buff, Blue and Gold Partridge, but the Lavenders remain the most popular. The Silver Laced is the latest type to gain popularity, although the newer colorations often lack the compact and tilted breed standards of the classic show-scene birds. In the United States, Pekins are confusingly known as Cochin Bantams.

PENEDESENCA

TYPE: Large fowl • EGGS: Dark reddish brown
CHARACTER: Alert, energetic forager

A rare, exotic and bouncy layer
of chocolate-bloomed eggs

ALL MEDITERRANEAN HENS are flighty and light-bodied, and they all lay white eggs, as indicated by their white earlobes. The Penedesenca hens, however, reared in the Catalan region of Spain since the 1980s, lay beautiful chocolate powder-coloured eggs that rival those of the French Marans hens, even though they display very clear white earlobes!

These birds really need to be ranging freely about the place to be contented, and have floppy combs, like the Leghorn hens, that bounce and wobble as the hens run and flap about. Their eggs are incredible, sometimes with a wonderful purplish bloom to them when the hens first come into lay, but the thrilling colour does fade as more eggs are laid across the season, and only certain hens will have these fabled genetics. This rare breed is not much seen outside Spain, with just one person, Sophie Macaulay, rearing them in Britain.

PLYMOUTH ROCK

TYPE: Large fowl and bantam • EGGS: Light brown
CHARACTER: Confident, good layers, hardy

A strikingly productive layer, deservingly favoured
by smallholders in her native United States

THE BARRED PLYMOUTH Rock is memorable among fowl, like a moving humbug in its black-and-white feathers. Buff Plymouth Rocks, a lovely Toblerone-packet yellow, are very fine layers among buff-coloured fowl, but they are unfortunately now quite hard to find.

Plymouth Rocks are good hens for the novice, being docile and productive layers of light brown eggs – about 200 a year. They are especially lovely in their bantam form, with a smart and tight-feathered appearance, and these are also good layers. They look similar to Barred Wyandotte bantams but can be told apart as they have a single, rather than a rose, comb.

A diet full of green grass will ensure your birds sport especially bright yellow legs. As for all chickens, you need to mow the grass regularly within a fenced area or moveable run, or get sheep or cows to graze it, to make it easier for the birds to peck at the blades.

POLAND

TYPE: Large fowl and bantam • EGGS: White
CHARACTER: Friendly once settled, vulnerable to bullying

Unforgettable hens in pom-pom bonnets of feathers,
these fancy fowl are surprisingly good layers

I OFTEN RECOMMEND Polands to newcomers to poultry keeping because they can become very tame, are surprisingly good layers, are decorative and they don't go broody. They have a wonderful appearance thanks to their head feathers, and also lay a plentiful supply of white eggs without ever seeming to consider sitting on them – broodiness is an annoyance to those starting out with chickens. Polands, however, must be given proper attention by keepers both novice and experienced as their profuse head feathers easily fall victim to mite attacks. Regular checking will prevent this, and they do keep themselves nice and tidy if they are given access to dry soil for dust-bathing.

It is important not to scare Polands. Their crests limit their eyesight quite badly – they can only see straight ahead and a little bit downwards – and if startled will become understandably aloof towards people. This hindrance of their feathered crowns means that they are liable to be bullied by other more dominant breeds if they are kept penned rather than free range. But handle them often as chicks and they will become incredibly tame and even make a brilliant choice of breed for considerate children.

The majority of ready-made hen houses with attached roofs on their runs provide perfect accommodation for Polands. They have very thin skulls under their feathered hats, so need to be able to keep dry and sheltered, especially during the wet and cold of winter. Drinkers that have a thin gulley, placed on a tall, heavy plant pot or bucket, will help them to keep their crests dry when they take a drink.

The breed is often mistakenly called 'Polish', but their

correct title is Poland. Despite their name, their ancestry is uncertain, apart from them coming from Europe, and a number of oil paintings depict them from the sixteenth century onwards. The cockerels are incredibly loud crowers, which is unfortunate, but they do look quite stunning and the breed comes in an entire carnival of colours; the warm-honey Chamois is my favourite. A number of strains are bearded, with feathers under the beak, making them even more of an extravagant sight as they wander about the garden.

The large fowl versions of Poland are very hard to find now, but the bantams remain hugely popular. Aside from the need to keep their feathers in good order, they are a robust fowl and I have never known any other breed's eggs being as reliably fertile when they arrive by post. The chicks mature very quickly too.

RHODE ISLAND RED

TYPE: Large fowl and bantam • EGGS: Light brown
CHARACTER: Confident, docile, hardy

Robust and steady hens with treacle-black feathers, famous for their egg-laying ability and ancestors of Britain's battery birds

MY NAN MIN well remembered her mother keeping 'Rhodie Island Reds', as she called them, for their eggs. 'Great big fowl,' she used to say. Bred by farmers of the Narragansett Bay in the state of Rhode Island in the 1830s, the breed's brown eggs were favoured in Britain, where they were seen as wholesome, like brown bread, whereas American culture preferred white eggs, seeing them as cleaner. Its impressive laying ability has been bred into the humble hybrid battery hen, a smaller bird who, alas, could be crammed into cages with more ease than her larger Rhode Island ancestors.

Rhode Island Reds are to this day an easy-going, productive and hardy breed. The name is often mistakenly, if not deliberately, mis-applied to the readily available hybrid birds. True Rhode Island Reds are large hens with very dark brown, almost mahogany-red feathers.

Before hybrids became the best of layers in battery cages, Rhode Island Red cockerels would have been run with Light Sussex hens to create good commercial layers.

ROSECOMB

TYPE: True bantam • EGGS: White
CHARACTER: Active, alert, confident

A spirited, dapper little flapper of a bantam
with iconic earlobes like delicate soft
peaks of blobbed meringue

THE ROSECOMB, BRED in Britain during the fifteenth century, is what is known as a 'true bantam'. These very small old breeds of chicken have no large fowl counterpart; the bantams of large fowl are known correctly as 'miniatures' rather than bantams.

Quick on their feet, flighty and noisy, Rosecombs are busy chickens made for farmyards and rural places where the males can crow to their heart's content and guard their hens without upsetting the neighbours. I once came across a bachelor flock of them on a stable gate, a wonderful sight of animated black cockerels in crowing competition, stretching and arching their necks proudly – and creating a din that soon led to complaints to the local council.

The breed's glistening comb begins from the bird's beak as a fat, raspberry-pink sea-urchin affair and finishes at the other end as a tapering red spike. The hens lay just a few small cream eggs in springtime. You should take these precious eggs to a trusted broody Bantam hen to incubate and rear the chicks. A reliable foster hen will do a better job than a female Rosecomb, who is often aloof as a mother.

SCOTS DUMPY

TYPE: Large fowl and bantam • EGGS: Cream
CHARACTER: Active, friendly, hardy (but should avoid mud)

Hero of old Scottish battles, a toddling short-legged fowl with a wonderful name

PERHAPS THE EMBLEM of Scotland should be a Scots Dumpy rather than a Scots Thistle, if we are to believe the folklore. Once known as the 'time-clock bird', Scots Dumpy cockerels, with their acute hearing, were said to have been carried with the Scottish armies of old because they immediately crow if awakened, be it night or day, so act as dependable little watch-dogs to warn of potential night-time attacks. One story tells of Norsemen trying to outwit the cockerels, whom they hated with a passion, by approaching the Scottish camp barefoot and therefore quietly. But they had not considered the thorns of the Scottish thistles! Howling and cursing soldiers, with thorn-harpooned feet, quickly awoke the Scots Dumpies, who crowed in alarm. The kilted Celts won the battle and the attackers were well and truly thwarted. Ever since, the Scottish Thistle has been the emblem of Scotland, but surely the Scots Dumpy deserves proper recognition too? Instead they became a rare breed and had almost vanished by the middle of the twentieth century. Lady Violet Carnegie was there saviour. She took a flock with her to Kenya and their descendants were later reimported.

Usually seen in either Cuckoo or Black feathers, Dumpies have deep bodies and would traditionally have been required both to lay well and give a good carcass. The hens also prove to be very good broody hens.

In terms of care, you need to keep the grass well mown to assist with the Dumpies' waddling. The Renaissance naturalist Aldrovandi wrote in 1598 of the breed as 'creeping over the earth' and the legs of pure-breed Dumpies should be no more than two inches long. A number of Scots Dumpy chicks will have legs that are too long. These should not be bred with in order to keep the iconic breed standard. Some eggs will not hatch at all due to the Dumpy's dwarfism, which can lead to what is known by the sad term 'dead-in-the-shell': when the chicks are poised to hatch but alas just don't bother to do so and die on their twenty-first day. (With other breeds, dead-in-the-shell is usually the fault of there being too much water in the incubator. When humidity is too high during incubation, the eggs' air sacks are unable to increase in size enough for the chicks to breathe within the egg and therefore hatch successfully.)

If you want a challenge, then take an interest in the Scots Dumpy. Once adult, they are hardy birds and deserve to be kept going, both for the breed's character and its thrilling wartime history.

SCOTS GREY

TYPE: Large fowl and bantam • EGGS: Cream
CHARACTER: Confident, energetic forager, hardy

An upstanding chequered, hardy bird with smart plumage that complements the native Scottish tartan

THIS OLD BREED, originating in the sixteenth century, was once known as the crofter's or cottager's fowl. A hardy, tightly feathered hen, strongly built for ranging about in all but the most inclement weathers, she lays large, cream-coloured eggs in good numbers and for many years.

They are striking birds, carrying barred feathers of steel-grey and black, with the black forming a V on their neck feathers. This type of feather is the breed's only coloration. Their upright stance is due to the Scots Grey having some game bird ancestry. Many believe that they are cousins to the Scots Dumpies, perhaps due to their similar tartan-patterned feathers.

Unsurprisingly, given the shortness of a Scottish summer, the chicks of Scots Greys rear quickly, and soon gain the characteristic black-and-white barred wing feathers. Scots Grey hens will go broody but prefer to be left to their own devices for this task and will sit in a place of their choosing, which isn't often a wise decision. A broody hen of any breed should always be moved to the most suitable, private and safe nest within a large hutch or separate hen house. Do this in the depth of the night when she is dozy so that by morning she awakens peacefully in her new nest, as if nothing had happened. Have some dummy eggs in the new nest that she can sit on for a day or so, to make sure she settles in, before replacing them with the eggs to be hatched. Always write in a diary the date that the hen starts to sit on her eggs so you know when they will hatch, which should be in twenty-one days' time.

SEBRIGHT

TYPE: True bantam • EGGS: White
CHARACTER: Active, alert, vulnerable to cold and stress

Golden Sebright

Tiny and constantly poised, with a striking fishscale dress of feathers

NO CHICKEN IS more perfect in both its plumes and outline than these smart little bantams. They demand your instant attention, with feathers patterned like the icing on top of millefeuille cakes. The breed was developed by Sir John Sebright in the early nineteenth century and he spent more than two decades crossing Golden Nankin Bantams, Rosecombs, Hamburgs, henny Game Fowl and Polands, choosing only the smallest birds of these crossings to eventually perfect the divine Silver Sebright as a true bantam.

Several old English breeds once used for cock fighting are part of the Sebright's genetic make-up and this explains why some of the cockerels suffer badly from Little Man Syndrome. Given the chance, they will delight in sparring with their keeper's boots. The hens can also be fierce in their temper towards one another. Newcomers must be introduced to the flock gradually and with great care.

Sebrights are ideal for quiet gardens or stable blocks, very happily flapping and scratching about; their small feet cause little damage. The hens and cocks form strong pair bonds and the hen's small eggs are like white marbles, with little more than a few dozen laid each year. They will often sit and then do a good job of rearing their mice-like yellow-and-grey chicks, with some assistance from you; keeping them in a rat-proof ark will protect them from rodents and crows when they are very small. For the best chance of successful hatches, fertile eggs are usually laid once the warmth of spring properly takes hold from winter. Vitamin D from sunshine is a great aid to the libido of chickens!

The Sebright is almost unique among British breeds of

bantams in being what is known as 'hen-feathered'. This is when all of the cockerel's plumage, rather than being pointed, is rounded like that of the hens. The males don't have saddle feathers, but the sexing of the growing chicks isn't too hard as the young cockerels soon show up as being redder in their tiny combs than their sister hens. The females show very little in the way of combs or wattles until they are almost fully grown. Good hens of the breed will have a dark mulberry rather than red comb.

As well as the famous Silver Sebrights, the breed also comes in Gold and Chamois colours. The latter is the most recently created and a little washed-out for my taste, but the Gold is striking. The Silver Laced Sebright passed on its genetics for striking attire to the Silver Laced Wyandotte. As a result of this feathered ancestry, Wyandottes were once called American Sebrights. If you really fall for laced feathers but are new to chicken keeping, then first keep the more robust Wyandottes in order to learn the basics as Sebrights are fragile, even though they are also very active little birds.

Unfortunately, these little bantams are very prone to the Marek's herpes virus that tends to strike growing birds, fatally attacking nerves in their legs and wings. Affected birds suddenly appear drunk and then crippled. Only vaccination can assure this does not happen, which isn't usually possible for small fanciers raising just a few chicks a year. Stress-free and clean conditions are an aid in beating Marek's disease, but its dreadful symptoms can develop as if out of nowhere.

Silver Sebright

SICILIAN BUTTERCUP

TYPE: Large fowl and bantam • EGGS: White
CHARACTER: Alert, active, vulnerable to cold

Beautiful hens both in name and look, with
dashing and alert heads carrying a cup-like
comb with a castellated rim

SICILIAN BUTTERCUPS ARE very scarce in Britain but more popular in the United States. This may be due to some states having more reliably mild winters. The bird's iconic comb can suffer from frostbite in cold winters and needs to be smeared with Vaseline to protect it from damage. This task is more easily said than done and is best attempted at night, when the birds are calmer.

Buttercups arrived in America in 1860, having been noticed at a Sicilian port by a New England sea captain. They sailed to Boston in a cage among crates of figs and oranges. The hens clearly had good sea legs as they laid very well during the voyage. This charmed the crew, which was lucky, as the original purpose of putting them on the boat was for the birds to be eaten.

Despite being native to Sicily, the very first Buttercup fowls arrived in Britain from America, imported in 1912 thanks to the efforts of a Mrs Colbeck of West Yorkshire. A British breed club was formed and had around a hundred members after the First World War, with the breed being noted as a good layer of white eggs. But by the 1920s, the British interest in the breed alas declined, due to more productive Mediterranean breeds such as the Leghorns taking centre stage; a sad shift, given the beauty of these spirited hens.

I've only come across the breed once, when I photographed them. The comb really is reminiscent of a stag's impressive antlers, especially in the headgear of the cockerel.

SILKIE

TYPE: Large fowl and bantam • EGGS: White
CHARACTER: Broody, docile, friendly, vulnerable to bullying

**Maternal teddy bears of the hen world
with candyfloss plumage**

NO BREED PROVOKES a more Marmite reaction of affection or distaste than the Silkie. With their fur-like feathers, they look like impersonators of angora rabbits.

This is undoubtedly a breed of Asian descent, but exactly where is hard to pinpoint. They first arrived in Britain from India in 1850, but they have a long lineage before that. The Venetian traveller Marco Polo described them in the thirteenth century as 'hens which have hair like cats, are black and lay the best of eggs', and the Renaissance naturalist Aldrovandi wrote that they were 'white as snow with wool like sheep'. The breed's distinctive feathers are due to a lack of structural webbing. Even the Silkies' flight feathers are completely useless and they usually won't attempt to flutter, even over low fences. But you should still offer them a low perch for roosting at night and, like all breeds, they'll dust-bathe to keep themselves in good fettle.

This docile breed is liable to be bullied by more robust hens and is better kept apart. The hens have a strong wish to sit on eggs and broodiness will be a bore if you don't want eggs to be hatched. However, Silkies from show lines tend to sit much less often. These show birds have very good cushions (tail feathers) and a neat crest that resembles a powder puff. A Silkie out in the rain quickly looks like a cold and dirty dishcloth, but they usually learn to seek shelter quickly during summer downpours. In the winter, they need to be offered a cosy deep litter run with a roof.

When they aren't sitting, Silkies are active little hens and lay small white eggs that are the perfect size for children's breakfasts. They come in White, Black, Cuckoo, Blue, Partridge and Gold, all with wonderful turquoise earlobes that look like they are wearing sapphire earrings. The Gold, I think, is the most teddy-like in its appearance, whilst the most familiar kind, the White Silkie, reminds me of my Grandmar Sheila's white perm. My mum doesn't like the look of the cockerels, and I must admit that the comb does resemble a beetroot-coloured burst brain!

The Silkie has always been a small breed, but there is also a miniature bantam version. These are so tiny that they are no good for sitting on anyone else's eggs apart from their own. Be aware that Silkies seem to stress easily and get ill, especially when being moved, so buy birds from reputable sources to avoid new birds developing any colds. Good fanciers and large hatcheries will vaccinate them against the fatal Marek's disease, a virus for which there is no cure, only prevention by vaccination of day-old chicks.

SILVERUDD

TYPE: Large fowl and bantam • EGGS: Olive
CHARACTER: Alert, energetic forager

**Exclusive and easy-to-please layers
of olive-coloured eggs**

SILVERUDDS ARE THE only recognized pure breed that lays an olive-coloured egg. There is now a fashion for such unusual eggs, both olive-coloured or olive-blushed, spurred on by the popularity of blue eggs, and you also get these curiously shaded eggs from Olive Eggers, produced by crossing blue-egg-laying Cream Legbar or Araucana cockerels with Copper Marans hens. Martin Silverudd of Sweden gave the breed its name after he developed it in the 1940s by crossing the New Hampshire Red and Rhode Island Red with the coloured-egg genetics of the Cream Legbar.

This is a wonderful, rare-breed addition to a flock of hens that you can keep for the thrill of creating boxes of colourful eggs. The medium-sized hen has a confident, active nature and lays around 200 of her distinctive eggs in her first season. The breed comes in Splash and Blue feather colours. I think the Splash is the most beautiful, with marbled feathers unevenly sprinkled with tips of dark grey on a soft base of grey plumage.

SPANISH

TYPE: Large fowl and bantam • EGGS: White
CHARACTER: Docile, friendly, vulnerable to poor weather

Impressive *Phantom of the Opera* cockerels and
hens with faces like a Tudor queen's white make-up

A SPANISH FOWL, thought to be native to Castile. Their faces are too punk-rocker style for me, but as my great-aunt Iris would say: 'Each to their own.' The white on the face should extend above the eyebrows and also go below the wattles, in the case of the cockerel. When I sent my drawings to the great US poultryman Mr P. Allen Smith, he kindly said my cockerel would be a blue-ribbon winner (the poultry standard's first prize) for his excellent comb.

Spanish hens are impressive layers of white eggs and flocks were kept for this reason in the late nineteenth century, but they fell from favour, succeeded by other Mediterranean breeds, most notably Leghorns. This is a friendly, active breed, but their chicks are delicate little things, and their white faces need shelter from the elements to prevent any disfiguring blisters, although they cope well with hot climates.

SULTAN

TYPE: Large fowl and bantam • EGGS: White
CHARACTER: Confident, tame

**A decorative rare breed with looks suited
to strutting about a palace garden**

TURKEY IS THOUGHT to be the country of origin of this abundantly snow-white-feathered breed. The name is derived from the Turkish name of Sarsi-Tavuk, which translates as 'fowls of the palace', or Seri-Tavuk meaning 'sultan's fowl', from when they were kept in the gardens of the Ottoman sultanate.

They are purely ornamental birds, being not much good as layers or for eating, of a very calm disposition and with fine feathers that suffer in cold and wet climates. To look at a hen from afar, one could be forgiven for thinking she is some sort of white feathered-legged Marilyn Monroe dress.

The first Sultans reached Britain from Constantinople (formerly Istanbul) in 1854, given to a Miss Elizabeth Watts of Hampstead, but just five birds arrived and they had suffered badly on their boat journey. Miss Watts wrote that they had matted, dirty feathers and that she then struggled to rear enough birds from such sorry, runty souls. The small genetic pool of this breed has prevented Sultans ever becoming well established. In the nineteenth century, they were crossbred with White Cochins to try and improve their vigour and this has helped them to survive to this day, albeit in tiny numbers. Researching Sultans, however, I was pleased to find that they have a strong following in their native land, with several Instagram accounts of fanciers that show off very healthy-looking birds, not only in the familiar White but also in Black and Blue.

SUMATRA

TYPE: Large fowl and bantam • EGGS: White
CHARACTER: Tame if handled often

Incredibly ornamental yet surprisingly independent
(in the right setting), with a strut you won't
forget in a hurry

WITH THEIR SMALL bodies and large wings, Sumatra fowl are said to be capable of flying between the neighbouring Indonesian islands of Sumatra and Java. More like pheasants than chickens in their apparel, these mystical birds look like living Philip Treacy hats, designed for his beloved Grace Jones. Hens and cocks both have the most incredible oil-black green sheen, known by fanciers as 'gypsy'.

The birds will be completely wild unless they are hand-reared and handled often as adults, in which case they become very amenable. The poised hens lay little more than a hundred eggs in their first year of coming into lay and will sit and rear their own chicks superbly.

The cockerel's tail feathers become apparent within the first few months after hatching and get better with age, as with male peacocks. In Indonesia, the cockerels were primarily kept as fighting birds, but the breed is docile compared to other such birds and not bloodthirsty, unless provoked by unfamiliar, imposing males.

SUSSEX

TYPE: Large fowl and bantam • EGGS: Cream
CHARACTER: Broody, confident, docile, hardy

Docile and easily pleased, a broad and stalwart hen
ideal for both garden hen run and allotment

THESE BIRDS WERE first bred in the 1850s, with their breed club formed in 1903. It is the Light Sussex hen that many people know of as the Sussex, with its iconic nun-like collar of black hackle feathers against pure white. These roly-poly puddings of hens are traditionally good layers of light brown eggs and have a body that makes for a hearty casserole if you are inclined to cull old hens once they don't earn their keep. This was the backyard fashion of keeping them during the Second World War, when they became very popular. The hens, in both their large fowl and bantam versions, make excellent, calm-tempered broody hens.

The Light Sussex outshines the more shy colours of her sisters, the Speckled, the Buff and the seldom seen Silver. The Buff Sussex is of a good rich cornflake orange, especially so in the cockerel's plumage. Curiously you don't see many of them, despite the popularity of the Buff Orpington, which has the same coloured plumage. The Speckled deserves to be kept far more, a soft-natured large hen of feathers that are of a black-treacle and red-mahogany base speckled with white tipping that increases beautifully as the hens get older with each moult. I hope to have a flock of the Speckled myself one day.

These other colours mature much more slowly than the Light Sussex and need more pampering as chicks, and indeed protection from their more quick-to-mature Light Sussex cousins, who are liable to become bullies to them and to other slower-growing breeds if there isn't plenty of space.

If I were to keep the Light, I would seek out a good balanced strain that contains both show and utility genes.

Some of those bred purely for laying lack the lovely heaviness of the show strains, whereas the show strains lack the laying vigour of the utility birds. The late Fred Hams, a renowned poultryman, once had an excellent strain of Light Sussex that lived up to both desirable aspects with gusto: good looks, a good amount of white breast meat and laying vigour.

Beware the inferior Sussex hybrid hens, who are much slimmer and often more hawkish-looking than the pure-breed originals. My neighbour Charlotte has been unfortunate enough to buy two of these hens: aimlessly dull birds the colour of a baby's disposable nappy and not a patch on birds of pure Light Sussex descent.

Buff Sussex

Speckled Sussex

SWEDISH FLOWER

TYPE: Large fowl and bantam • EGGS: Cream
CHARACTER: Confident, hardy

Strikingly robust, with feathers uniquely
tipped in a kaleidoscope of colours

THESE PRETTY AND laid-back hens are splattered with varying degrees of white tips, known as flowers, mottled against a lovely marzipan medley of brown, orange, lavender, black and grey feathers. A flock of these hens is an especially wonderful sight, as are the cockerels; some of them have tufted head crests too.

This is an old landrace breed, which means the birds share the same character, having naturally adapted in isolation to suit their particular place, rather than being selectively bred. As such, they are very hardy against the cold, provided they are kept well-fed.

Swedish Flower hens became very rare in the 1970s in their native Sweden, but interest has grown after they were exported to America. A few breeders have attempted to create some different strains to ensure a more guaranteed feather pattern, such as one known as the Snow Leopard, but by and large the feather-flower patterns are a lottery.

THURINGIAN

TYPE: Large fowl and bantam • EGGS: White
CHARACTER: Active, confident, friendly

A pretty, exotic and friendly hen that looks like a muse for Old Master oil paintings of fowl

AFFECTIONATELY KNOWN AS 'chubby cheeks' in its native region, the Thuringian Forest in central Germany, the breed's prominent feathered beard hides its red wattles. This old, hardy and beautiful breed, recorded as far back as 1793, only reached Britain in 2000. Bevere Rare Breeds maintains a breeding flock and is the only place I have come across that rears Thuringians in the UK.

They are very active and busy, with friendly personalities, and quickly grow in confidence. The elegant, small hens are fair layers of white eggs, and they don't readily go broody, making them superbly suited as little pet hens who would be happy in a modestly sized garden coop, with plumage that complements your flower beds. A flock of them at large in a farmyard or garden, perhaps mixed with crested Polands, would be quite the show.

VORWERK

TYPE: Large fowl and bantam • EGGS: Cream
CHARACTER: Active, confident, hardy

**A dapper, ginger biscuit-coloured hen that's
an economical and hardy layer for the smallholder**

THIS GERMAN BREED was created in 1900 by Oskar Vorwerk, whose aim was partly to develop a hen of darker plumage that wouldn't show the dirt on her feathers. Vorwerks almost vanished after the Second World War when other breeds beat them in terms of egg production. They are now a rare breed – sadly, because they make little fuss, are good layers of cream-coloured eggs, look charming and are well worth keeping in a large garden coop or smallholding.

The breed's deep orange feathers get better with age. The males are the same colour as the hens, with plumage that verges towards a terracotta red in their saddle feathers, and are noted for being extra-tolerant of other cockerels. The black and golden-brown chicks mature and feather up very quickly, with the young cockerel's comb showing quite early on, and looking like a little scrap of uncooked bacon!

WELSUMMER

TYPE: Large fowl and bantam • EGGS: Terracotta
CHARACTER: Alert, active, hardy

Well-dressed layers of the most thrilling
brown eggs, and cockerels who could model
for the Kellogg's Cornflakes packet

ORIGINALLY BRED IN Holland, Welsummers are famed for their eggs, which many consider to be the most beautiful of all the colours: an orange-terracotta shade of brown. This shell colour was at first believed to be a trick because the dark pigmentation disappears if the shells are washed vigorously, but it is entirely natural.

The Welsummer is a striking, classic story-book farmyard hen to look at, running around on yellow legs, with every shade of plummy brown and chestnut upon her body and shimmering orange-and-gold neck feathers that are finely striped with black. The chicks are gorgeous bundles with flecks of Elizabeth Taylor-worthy Queen Cleopatra eyeliner.

These are active birds and need to have a free-ranging life of foraging or be kept in large hen runs. The breed can suffer from shyness. I once wanted to photograph a very beautiful flock, but the hens behaved dizzily and wouldn't come out of the hen house, despite my best efforts of crouching down and moving slowly. The bravest of the lot acted as a look-out, peeking out from the pop-hole to see if the stranger-with-a-camera had gone yet. Deborah Devonshire charmingly described them as a breed that 'faded into the shadows of the farmyard'.

But don't let my comments on their shyness put you off! Buy young girls, tame them with much chattering and many tasty titbits, and they will quickly overcome their nerves.

WYANDOTTE

TYPE: Large fowl and bantam • EGGS: Very light brown
CHARACTER: Broody, confident, docile, hardy

Blue-Laced Red Wyandotte

Surely the United States' finest offering to poultry breeds, in laced feather colours to suit every taste

Gold Partridge Wyandotte Bantams

THE WYANDOTTE'S COMPACT body makes the birds of this robust and hardy breed look like they are wearing a thick puffer jacket of feathers. Originally from the US and named after the indigenous Wyandot people, they were created by crossing many breeds. There was a lack of agreement on the correct type of comb they should have, and this meant the Wyandotte was not admitted to the recognized list of American poultry breeds until 1883, after it was decided that this should be its now iconic little flat-topped rose comb.

There are so many kinds of Wyandotte colour that choosing your favourite is like viewing gorgeous, expensive fabric samples, and will require careful consideration. The lacing is down to the influence of the little British Sebright bantam, and the first standardized colour was the divine Silver Laced Wyandotte. I especially love this coloration when they are little chicks and their black-and-grey markings make me think of little fledgling owls.

The Gold Laced is modestly striking, in a deep mahogany and the sort of orange you see in the dying embers of a fire, while the Partridge has the most perfect feathers of them all. Perhaps needing the most dedication, in terms of breeding them to perfection, yet the most in demand, is the Blue-Laced Red Wyandotte, a more recently created variety. Each of its breast feathers is a varnished deep-red cinnamon, laced with slate-blue edging.

The Wyandotte is a bonny, happy-go-lucky and robust breed that almost always has a bright face even when moulting. They carry themselves about busily on their bright yellow legs but will be content within hen runs and fit the bill for a family wanting a regular egg supply from friendly hens that are calm and beautiful, but not too fancy-plumed. The bantams are more popular than the large fowl, and the latter urgently need more enthusiastic keepers if they are not to be lost.

Wyandottes make fine broody hens, which can be a bore if this isn't needed, but they don't sulk too badly if sitting cannot be accommodated and you need to take them off their eggs. Put such broody hens in what is called the 'sin bin': a sturdy wire cat-carrier with food and water but no bedding. Place this somewhere safe but busy, so that the hen can't settle, usually for at least ten days. The wire floor may seem stark, but the point is to cool the broody hen down as quickly as possible so that she forgets the idea of wanting to sit on eggs and returns to her usual self before losing too much weight through endless and pointless sitting.

Silver Laced Wyandotte

KEEPING HENS

HEN HOUSE

A hen house should always feel cosy and airy, somewhere you would consider sleeping if you had been locked out of your own house! Plastic hen houses are more popular these days than their traditional wooden counterparts, with good designs giving the birds reassuring insulation against extreme cold and heat. I, however, find them visually poor and much prefer the aspect that a large, sturdy shed offers instead. My favourite garden hen houses are those designed by the Domestic Fowl Trust. Look to these for inspiration if you decide to build your own models.

Blood-sucking parasitic red mites thrive in dirty hen houses, where they emerge at night to plague the roosting hens. They sense body heat, so if you return from egg collecting or cleaning-out duties itching then you know you have a bad infestation. The ease of cleaning plastic hen houses is a point in their favour, but wooden housing can be kept mite-free with a little effort. Simply paint the interior with a lime-based paint – a white paint will make the hens look especially chic. Replacing roofing felt with corrugated Onduline sheeting also helps deter mites.

Inside the hen house, all the perches should be at the same height, so the hens don't argue over who gets the highest one. Ensure the perches can easily be removed so that you can inspect them for mites and dust them regularly with food-grade diatomaceous earth, which helps protect against mites and other parasites – horrible creatures – before replacing them.

Nesting boxes should be set up lower than the perches so the hens don't sleep in them. One nesting box per three hens is ample; put them in the darkest spot, directly under the window. Furnish them with wood shavings or chopped straw – never hay or whole straw as these can harbour mites, absorb moisture poorly and become mouldy – and consider deploying fake eggs to show hens that are coming into lay where to lay them. A cockerel will inspect the nesting boxes and cluck encouragingly if all is present and correct.

Dirty eggs are often due to the hen's feet being muddy, so an old hessian doormat placed at the entrance to the nesting boxes will help keep them clean.

The hen house should be cleaned out often: I remove their nighttime droppings each morning, treating the house as if it were a horse's stable. Every fortnight you should give the hen house a deep clean at dawn. A paint scraper, dust mask and dustpan are helpful tools, as is an old, robust hoover – if there is an electricity supply to hand its nozzle will work wonders.

Ventilation is very important; there should be a flow that encourages the air to pass over rather than under the hens as they roost. An under-draft is not good for the birds.

HEN RUN

A fruit cage will make for a good-sized, sheltered and secure hen run, so long as its netted sides and roof are reinforced with thick galvanized wire that a fox cannot chew through. Place the run alongside the shelter of a hedge, somewhere that the sun will shine on for part of the day. Hens will not thrive in damp, boggy or dark places.

It is wise to connect the hen house to a secure run. The flock can then be allowed out to free range when you are around and about to keep an eye on them. Shutting the hens' personal hen house door (the pop hole) at dusk is essential to avoid raids by foxes, badgers, mink and rats. A battery timer, known as a chicken guard, will open the pop hole at dawn and close it at dusk: provided you keep it timed correctly, this is an invaluable ally in the fight against villains and modern life's requirement of often being away from home. Dawn fox raids are common, so I don't like to let my hens out before 8 a.m.; the hens know to go to bed before dark themselves, so the chicken guard is timed to close just after dusk.

Moveable poultry arks are an option if you have large, flat areas of pasture. It is grass, after all, that makes eggs taste delicious. To keep the grass around the hen run and house viable, mud mats are well worth investing in as otherwise your footsteps will quickly create a muddy surround. Avoid placing paving slabs down as rats will often burrow under them.

Around the entire perimeter of the hen run, be it moveable

or static, a four-foot wide "skirt" of 12-gauge galvanized wire mesh should be pegged down tightly against the run's base. Foxes don't have the intelligence to reverse and dig a tunnel under the skirt.

When it comes to keeping the litter of a hen run sweet, there is no excuse for it to look and smell squalid. If the roof of the run is covered to keep it dry – which is recommended to protect against bird flu – you can create a good deep litter bed by combining chopped straw with dry leaves and fine wood chippings gathered in autumn. This can all be removed and replaced every few months and it will make a very nutritious mulch for the garden.

DUST BATH

Dust-bathing is often overlooked but this behaviour ensures healthy and beautiful hens; they will take great pleasure in bathing in a mix of dry, fine compost or earth, kiln sand and wood ash. A deep drawer placed somewhere dry, such as under a large garden table, will see much use by the whole flock, while an old greenhouse with its dry soil, a few of its panes missing and the door kept open will become a spa.

FEED AND WATER

Hens have a storage organ known as the crop that expands like a balloon as they feed. It allows them to digest their feed overnight, which is why hens produce a lot of their manure as they are roosting. Each hen will eat about a mug's worth of feed per day – they will quickly tell you if they are hungry. A compound layer's pellet ration will suit most laying hens but more fancy breeds will benefit from being on a breeder's or grower's pellet that has higher levels of protein. Buy the best quality feed you can, ideally one that is soya free.

Ab-lib feeding is best so the hens can help themselves to feed throughout the day. A little corn can be given in the winter towards dusk as a treat. If you choose – unwisely – not to invest in a rat-proof feeder, then you should definitely opt for hanging feeders instead. This will stop hens spilling too much feed onto the ground, thus attracting vermin.

Hens drink an average of 300 ml of water per day so fresh, clean water must always be available. Drinkers modelled on a nipple-pecking design will ensure clean water for the hens and can be hung up alongside feeders, but unfortunately they often look awful in comparison to a lovely antique galvanized drinker. A number of plastic drinkers have to be tipped upside down to unscrew and refill them, which is an annoying art form to master. Scrub drinkers weekly to keep them free from algae and scum.

THE PECKING ORDER

Chicken society starts to take shape within a week of chicks hatching, which means you should always work out in advance how many hens you wish to keep so that you buy them all together. If you intend on keeping a mix of different breeds, then they need to have been reared together to avoid what can be horrific bullying as the stronger characters of more dominant breeds really show their true and horrid colours.

The pecking order has a head hen who will always command respect. She'll be given precedence by everyone beneath, right from her second-in-command down to the lowly bottom hen, who ekes out a tough but manageable existence. Plenty of space will see all the hens get along each day. A cockerel can help manage the dynamics between hens, but he must be made to realize that you are the boss of him by regular handling.

The introduction of new birds to an established flock must be done slowly. Young hens must be older than eighteen weeks so that they can stand up for themselves. Let the flock get used to these newcomers for a week by letting them settle in a separate ark or large hutch. After this put them in with the main flock at night so that they all awaken together. A favourite bullying trait of resident hens is to push newcomers off the feed, so provide an extra feeder and give the flock lots of diversion. Usually after a fortnight, the bullying settles.

HOLDING AND BEFRIENDING CHICKENS

It is a mistake to smother a hen tightly against your body or to hold one in an upright fashion. You must be mindful of the pressure hands can put on a bird's ribs and lungs. Hens absolutely hate the feeling of friction ruffling their feathers, so their wiggling and flapping when uncomfortable is understandable. Never hold a hen upside down by her feet or attempt to grab her by one foot or one wing – this can cause hurt and injury.

Try to hold hens in what I call the bagpipe method. If you have a box of newly bought hens, slip your hand under the body of the bird and then clutch both of its legs gently but securely in the same hand. The hen's body is now supported by your forearm as you lift her out of the box facing towards your body. By supporting her body weight and holding her feet gently but securely in your hand she should relax and settle down. There is something about them feeling that their feet are secure that helps to steady them, even if they aren't used to being handled. You then also have a spare arm and hand to check her over.

Establish an evening routine with newly bought hens. Every evening, when they have gone to roost, go and chat to them. You can gradually progress from talking to stroking their chests in the direction of their feathers. After a week or so, once they are used to this, begin to handle each of them in turn using the bagpipe method, lifting them off and back onto their perches.

At dusk they will be far calmer than in the day. Never stress hens out by chasing them around; wait until they have gone to bed, then catch them should you need to.

Hens can see in full colour vision and certainly learn their keeper's voice and appearance. If you don't have a cockerel with your hens, then they will often crouch down when you are nearby and make a sudden movement. This can be a useful opportunity to pick a hen up – and when you put her back down, she will ruffle her feathers as if she has just had a liaison with a cockerel!

LAYING

Hens are not egg laying machines but, despite this, the first question you might be asked about your chickens is: 'Are they laying?' If you consider the amount of protein an egg contains and its calcium-rich shell, it is little wonder that all hens will take well-deserved breaks from laying them. It is remarkable that hens lay so many eggs when nature designed most female birds of other species to lay a single clutch of eggs each spring. You don't need the presence of a cockerel for hens to lay eggs.

Young hens are often bought when they are eighteen weeks old, when they reach 'point of lay'. This is when the young hens begin to cluck rather than cheep, their faces blush and their combs and wattles begin to ripen too; they are experiencing puberty, preparing for the onset of laying.

The first few dozen eggs laid are known as pullet eggs. Supermarkets dismiss them as being too small to be worthy of selling but small eggs should never be snubbed. Whilst a box of bantam eggs might not impress at first glance, when you break one, their yolks are surprisingly large – they actually contain more yolk than egg white. Bantam eggs are often considered to have a richer flavour too.

Although commercial hybrids usually begin laying from eighteen weeks old, pure breeds often won't begin until they reach the age of twenty-one to twenty-eight weeks, especially if they are winter debutant pullets. The fancy feathered breeds with an apt 'too posh to push' reputation may not start laying until longer spring days set in. It is the length of daylight that is the biggest laying stimulant to hens.

Within a commercial system, hens are influenced to lay continually through exposure to artificial lighting – the conditions are those of an eternal spring – at least until their bodies eventually demand a reprieve. Hens at home on the other hand will naturally begin to stop laying as the shrunken daylight hours of winter take hold.

Hens will lay the most eggs in their first three years of life. Eggs that are laid in their fourth and fifth years will be noticeably bigger, but they'll be laid less frequently.

ODD EGGS AND THEIR CAUSES

Blood spots in the yolk are caused by a blood vessel on the yolk's surface rupturing as it forms. Despite the unsightly visual, they are harmless and often occur when the hens are in full lay.

Soft-shelled eggs are usually the consequence of a lack of suitable grit being provided to the hens. It may be a calcium absorption problem if soft shells are laid regularly. Liquid calcium can help.

Double yolks are the result of two yolks being released together from the hen's fallopian tube, which tends to occur when older hens come back into lay after a break.

HATCHING EGGS

Before hatching any eggs, you must heed the very wise saying: 'Don't hatch if you can't dispatch'. This refers to the spare cockerels that hatching eggs will almost always produce. Most broods of chicks will have a majority of males, which will be a problem for urban keepers and those without space. Once matured, young cocks begin crowing and hassling the hens greatly as their hormones rise. Your original cockerel, should you have one, will fight with these teenagers and he may become an outcast of the flock as younger blood outranks him.

Unfortunately, there is no easy answer to spare cockerels. Very few will be able to be rehoused to trusted homes, though it is still worth advertising them. You might choose to keep your cockerels as a separate bachelor flock. In such circumstances, without hens to fight over, they will form a male hierarchy.

If this is not possible then it will be necessary to imagine the many wonderful soups and casseroles a young cockerel's gamey carcass can provide. Culling any chickens must be done confidently and with a skilled technique. The pull down, flick up and twist method is clean and effective. Culling at night when the birds are roosting will cause the least amount of stress. Seeking experienced help is best.

It is sensible to increase your flock naturally by letting a hen brood eggs and rear chicks. Chicks reared by a hen are always superior in their intelligence because of their natural upbringing. It is common sense to hatch eggs from spring onwards as nature intended so that the young birds can enjoy the summer sunshine, rather then beginning their life in the harsh winter weather.

Broodiness will be a bore if you do not want to hatch any chicks. In such instances the broody hen must be dissuaded from her mission. The sin bin, a small dog crate with a wire cage floor placed somewhere safe but busy, is where an unwanted sitting hen will need to be housed while her body cools down and her hormones return to normal levels. Provide her with feed and water. After fourteen days of this necessary lodging she should have forgotten the idea and can be released.

You will know when you have a serious broody hen as she won't budge from the nest box and she will hoard eggs under her. If she cuddles your hand enthusiastically with her body, then you know she is very serious about sitting. She can't, however, be allowed to continue sitting in the main hen house. Here her eggs will get muddled with freshly laid ones and the other hens will get annoyed with her.

The broody hen, therefore, must have her own broody coop. A secure rabbit hutch is ideal. It is vital that it is rat-proof and placed somewhere quiet and sheltered. The nest should be formed of a generous amount of clean, dust-free straw.

Gently move the hen to her maternal abode at night. Moving her in the dark means she won't get flustered as she would do if she was moved by day. Have a trio of dummy eggs or hard-boiled eggs waiting for her and let her sit them for two days before giving her any precious hatching eggs. This way you can be sure she's still seriously broody.

If you don't have an adult cockerel with your own hens, then the hatching eggs will need to come from a respectable fancier. Eggs can be posted but their fertility is always negatively affected, so it is worth the effort to collect the eggs yourself. Rest the eggs for a day after their travels somewhere cool.

The clutch of hatching eggs must be given to the hen all at the same time so that they hatch together after twenty-one days. The hen will accept eggs of any colour! Take away the fake eggs as you give the hen the hatching eggs. It is important for you to encourage the hen to take a luncheon and loo break each day by lifting her off the nest.

You must make sure that your broody hen's nest is free from any mites whilst she sits. Often, two weeks into her sitting, I replace the original straw with a fresh nest whilst she is taking a break, and so long as it looks like it was when she left it, she won't notice this.

The hen must be able to take a dust bath during her breaks. Let her have a good run and peck around so that the other hens of her flock don't forget that she exists. This will be helpful when she begins to reintegrate back into the flock when her brooding is finished. Her breaks will usually take between twenty to forty minutes; never rush her back to her nest.

About seven days into incubation you should candle the eggs. Candling is when you shine a bright light – a mobile phone light is ideal – through the egg to see (hopefully) a spider impression of developing blood vessels from the chick's embryo. Doing this at night by the hen's nest is the best thing, rather than the risk of taking the eggs elsewhere and breaking them. If the egg is clear then discard them. Obviously dark brown eggs are tricky to candle due to their opacity but the air sack, if the egg is fertile, will nonetheless be visible at the fat end of the egg.

When the hen's eggs are due to hatch, don't disturb her. The hen needs peace and quiet as she hears and feels the chicks emerging from their eggs. Provide her with a shallow-lipped chick drinker by the nest and a little dish of chick crumbs too. The hen will be keen to eat the chick crumbs, but they'll do her no harm. Short, fresh grass and dappled sunshine will see the chicks thrive. If the hen has become a little pale-faced during sitting then mashed carrot, hard-boiled egg and cod liver oil will be good for her condition.

When the chicks are four weeks old, let the hen chaperone them into the main flock for protection. They'll learn a lot from their mother and life will be much easier for them this way. Eventually, the hen, usually when her chicks are eight weeks old, will decide she has had enough of motherhood. By then, the youngsters will know where to go to roost and understand their place in the pecking order. They will hang around in their own youth club!

MOULTING

From the age of two, hens will moult their feathers every year, ideally at the end of summer if they have some sense. Hens will always stop laying when they are moulting as they have to devote all their energy to growing new plumage as fast as they can. Sometimes hens undergo a hard moult, losing a large number of their feathers overnight to wake up looking bedraggled. Their new pin feathers emerge looking like little pen nibs. The hens will be very quiet during this time: try not to handle them much as it will be uncomfortable for them. To help the new feathers grow quickly you can mix in two tablespoons of cod liver oil to a mug of layer's pellets. Most moults take two to three months to complete.

HENS IN THE GARDEN

The happiest chickens are those that spend much of their day under a canopy of branches. Hens draw comfort from the shelter that small trees and shrubs provide.

Bantams are always a far better choice for the garden-conscious as they have smaller claws to scratch about with that do a little less damage, and their smaller droppings are less noticeable on the patio! Defend vegetable beds with ornamental barriers, old, rusted metal gates or fire guards as otherwise the hens won't be able to resist the loosely dug soil and tasty seedlings. Fencing will save a lot of annoyance! The nitrogen-rich manure from the hens, however, is the best mulch a vegetable garden could wish for.

All chickens will indulge in the pecking of luscious foliage, and herbaceous borders in particular are at risk when they awaken in the spring with their clumps of fresh shoots. A garden that has a backbone of roses, shrubs and woody herbs will cope well, likewise narcissus taste horrid so the spring display of these is always ignored by the flock.

Hens will spend a good amount of time pecking on the lawn. Hens will peck the tips of grass each day, but don't feed them lawn clippings as they won't be able to digest them easily and this can lead to a horrid, often fatal condition known as sour crop. Slugs and snails should not deliberately be fed to chickens either as they carry internal parasitic worms that adversely affect the health of hens. Hens in fact should be wormed once a year, like dogs and cats, by giving them a medicated feed containing Flubenvet for a week.

RATS

Rats are attracted by messy hen-keeping and by feed left lying around. They spread disease and, increasingly, will predate roosting hens. They have always been a menace when it comes to stealing eggs and killing young chicks, not to mention giving keepers the shudders when fleeing from the hen house. They must be avoided at all costs!

Investing in rat-proof feeders from the outset will save much trouble, as will raising the hen house off the ground. Always remember that rats like to feel their backs up against something. Any areas of clutter and mess near the hens should be avoided, and likewise do not compost nearby as rats love to burrow into a warm heap. Store feed securely in metal feed stores or lockable galvanized bins.

Some rat-proof feeders work by the hens treading on a peddle that opens the feeder, while others hang up and have feeding slots for beaks that close if reaching paws cling on them. You will need to teach the hens how to operate the peddle feeder, but once one learns, the others will follow. Feeders and ideally drinkers too should be placed under the cover of the hen run or the hen house to avoid attracting wild birds, which will often transfer lice and internal parasites to the hens. Pheasants are to be discouraged from visiting. Jackdaws steal eggs, but they can be outwitted by dangling a chain over the pop hole door – they won't like its dazzle!

LEGISLATION

It is now a legal requirement in the United Kingdom to register your chickens with the Department of Food and Rural Affairs due to increasingly widespread outbreaks of avian influenza. You will be notified when there is a risk of contagion in your area, which will sometimes entail a housing order, in which case your poultry must be kept penned under a covered run to avoid contact with wild birds until the risk has passed.

Under the 1950 British Allotment Act you are allowed to keep hens on your allotment, so long as the eggs are only for the use of the tenants and not used for business or profit.

It is worth checking the deeds of a property for poultry-keeping restrictions. The crowing of a cockerel will, alas, often result in noise complaints from neighbours, so housing cockerels in especially solid hen houses and not letting them out early is wise to help muffle their alarm clocks!

GLOSSARY

AUTO-SEXING Describes breeds where the males and females can be easily told apart as day-old chicks due to barring or striping on their fluff. The name of each such variety ends with the word 'bar', e.g. Cream Legbar.

BANTAMS Usually small-sized fowl miniatures of large fowl that are about a quarter of the weight of their large-breed counterparts. True bantams are much smaller, with some being smaller than pigeons.

BARRING Alternate stripes of dark and light across a feather (*see* Barred Plymouth Rock, Scots Grey and Legbar). Also used in reference to sex-linked barring gene (*see* auto-sexing).

BEARD Tuft of feathers under the beak (*see* Faverolles, Houdan and some varieties of Poland).

BOOTED Feathered shanks and toes (*see* Brahma and Cochin).

BROODINESS Behavioural pattern in hens wishing to incubate and brood a clutch of eggs. Such hens are reluctant to leave the nesting box.

CANDLING Method of inspecting the contents of an egg using a bright light in a dark place to see the embryo, usually done seven days into incubation.

CARRIAGE The attitude, bearing and style of a bird, especially its walk as it struts its stuff.

COCK A male bird aged over twelve months, or after its first moult.

COCKEREL A young male bird under a year of age, before it becomes fully mature and moults into its true adult feathers.

COMB Fleshy growth on top of the head varying in size and type. Healthy adult birds should display a comb of a good red complexion.

CONDITION General state of a bird's health. Good condition would be bright eyes, alert, with clean and well-preened feathers.

CREST Tuft of feathers on top of the head, also known as a 'top-knot' (*see* Faverolles, Houdan, Poland, Araucana, Legbar and Silkies). Known as a 'tassel' in Old English Game.

CUCKOO Indistinct and irregular bands similar to barring (*see* Cuckoo Marans).

CUSHION Mass of soft feathers on the hen's rump covering the root of the tail (*see* Cochin).

DOUBLE-LACED Two thin lines of black, one on the edge and one set in from the edge of the feather (*see* Cornish Game Hen).

DUAL-PURPOSE Describes a breed that both lays well and is good as a table bird. These are usually old, traditional breeds such as the Sussex.

DUST BATH An area of fine soil or sand in which the bird cleans its plumage and controls external parasites. It is essential to provide a dust bath for confined hens to keep them healthy.

EARLOBES Folds of skin hanging below the ears with variations in shape, size and colour, which can be purple, turquoise, cream, red or white. Hens with white earlobes usually lay white or blue eggs.

FANCIER Breeder of exhibition birds.

FEATHER-LEGGED Describes a fowl with feathers growing on their shanks and/or toes (*see* Brahma, Cochin and Faverolles).

GROWER A young bird from the age of six weeks to maturity.

GYPSY FACE Dark mulberry or purple facial skin.

HACKLES The long, narrow and pointed neck feathers and the saddle plumage of the male.

HARD FEATHER Close, tight feathering, as in game birds.

HEN Female bird after her first adult moult.

HENNY (OR HEAD-FEATHERED) A male bird without hackles and sickles (*see* Sebright Bantam).

LACING Dark single or double stripes around the edge of a feather, with the inner stripe narrower than the outer. (For single, *see* Andalusian, Wyandotte, Sebright; for double, *see* Cornish Game).

MILLEFLEUR The well-named thousand flowers colouration, which usually manifests as a combination of rich ruby red and marmalade orange feathers adorned with black and white tipping. There is also lemon millefleur colouration where the base feather colour is a light yellow, and you will sometimes encounter silver millefleur that is white and black.

MITES Blood-sucking poultry parasites, in the form of the red and northern fowl mite, that cause much discomfort, anaemia and eventually death to the affected birds. Regular cleaning and disinfecting of housing and the monthly powdering of birds with an anti-mite powder should be done to deter them.

MOONS Round spangles on feather tips.

MOSSY Confused or indistinct marking, usually considered by show breeders as a defect.

MOTTLED Marked with tips or spots of a different colour.

MUFFED With tufts or whiskers of feathers on each side of the face (*see* Faverolles, Houdan and some Polands and Araucanas).

PEA COMB Triple comb, resembling three divided small combs, the central one being the highest, and joined at the base (*see* Brahma).

PECKING ORDER The hierarchical order within a flock.

PENCILLED With two to four black lines, varying in thickness, usually following the contour of a feather (*see* Gold Pencilled Hamburg).

PEPPERING The sprinkling of a darker colour over a lighter background of feathers.

PIN FEATHER A growing feather that is not yet fully opened from the quill, resembling a Biro pen's ink tube and tender to the touch. Seen in growing chicks or adult moulting birds.

POINT OF LAY (POL) The period when a pullet is about to begin laying her first eggs – normally between sixteen and twenty weeks in commercial hybrids but more like twenty-four to thirty weeks in most pure breeds.

PULLET A young female of the current season's breeding, or between six weeks old and her start of lay.

ROOSTER Old English and American name for a cock.

ROSE COMB Broad comb with a flat top covered in small points and finishing in a spike, or leader.

RUST Patch of red-brown colour on the plumage of usually black-red females, also known as foxiness.

SADDLE Corresponding to the cushion in the female (see above), and covering the whole of the back of the male, reaching to the tail.

SICKLES The top pair of especially long curved tail feathers of a male bird (the shorter tail feathers are known as the lesser sickles). Sometimes used to refer to the tail coverts, which are the shortest tail feathers that cover the bases of the taller, more magnificent sickle feathers.

SIN BIN A wire cage placed somewhere secure but without comfortable bedding to cool down an unwanted broody hen so she no longer wishes to sit on eggs.

SMUTTING Brown feathers or brown marks within plumage that should be of a pure buff coloration, a fault seen in breeds like the Buff Orpington. Such birds are for home hen keeping – their personalities are just as wonderful – rather than being show quality.

SPANGLING A contrasting circular spot of colour at the end of each feather. In the case of the Poland it consists of broader lacing at the end of the feather.

SPLASH Irregular patches of dark grey or black feathers that interrupt the main body colour of a bird that is either white or a very light grey. Splash feathering should occur in all areas of the bird's body.

TRIO A male and two female chickens.

WATTLES – Fleshy appendages hanging either side of the beak and more pronounced in the cockerel.

FLOCKS FOR ALL OCCASIONS

LAYING BREEDS THAT THRIVE AT LIBERTY

- Andalusian
- Appenzeller Spitzhauben
- Araucana
- Derbyshire Redcap
- Hamburg
- Legbar
- Leghorn
- Old English Game
- Old English Pheasant Fowl
- Penedesenca
- Scots Grey
- Welsummer

GOOD BACKYARD AND ALLOTMENT LAYERS

- Burford Brown
- Dorking
- Lincolnshire Buff
- Marans
- Marsh Daisy
- New Hampshire Red
- Orpington
- Plymouth Rock
- Poland
- Rhode Island Red
- Sussex
- Wyandotte

PARTICULARLY TAME AND DOCILE

- Belgian Bantam
- Brahma
- Cochin
- Cornish Game
- Faverolles
- Orpington
- Pekin
- Poland
- Scots Dumpy
- Silkie
- Sultan

GARDEN ORNAMENTALS

- Belgian Bantams
- Brahma
- Cochin
- Dutch Bantam
- Japanese Bantam
- Pekin
- Poland
- Rosecomb
- Sebright
- Silkie
- Sultan

USEFUL BOOKS AND WEBSITES

BOOKS

Ian Kay, *Stairway to the Breeds*, Scribblers Publishing, 1997
Victoria Roberts, *Poultry for Anyone*, Whittet Books, 1998
Francine Raymond, *The Big Book of Garden Hens*, Kitchen Garden, 2001

BRITISH WEBSITES

poultryclub.org
rbst.org.uk
rarepoultrysociety.com
gbpoultry.com
legbarsofbroadway.co.uk
henkeepersassociation.co.uk

AMERICAN WEBSITES

amerpoultryassn.com
livestockconservancy.org
heritagepoultry.org
greenfirefarms.com
mcmurrayhatchery.com
marthastewart.com

ACKNOWLEDGEMENTS

With huge thanks to Sam Fulton, Olga Kominek and the team at Penguin, Samatha and her colleagues at the Chatsworth Farmyard, Hattie Ellis and Gordon Wise. And thank you to James Mackie for putting up with me as I insisted on writing and drawing into the night for many months to finish this book.

PARTICULAR BOOKS

UK | USA | Canada | Ireland | Australia
India | New Zealand | South Africa

Particular Books is part of the Penguin Random House group of companies
whose addresses can be found at global.penguinrandomhouse.com.

Penguin Random House UK
One Embassy Gardens, 8 Viaduct Gardens, London SW11 7BW

penguin.co.uk
global.penguinrandomhouse.com

First published in Great Britain by Particular Books 2025
001

Copyright © Arthur Parkinson, 2025

No part of this book may be used or reproduced in any manner for the
purpose of training artificial intelligence technologies or systems. In accordance
with Article 4(3) of the DSM Directive 2019/790, Penguin Random House
expressly reserves this work from the text and data mining exception.

The moral right of the author has been asserted

Set in Plantin MT Pro 9.6/13pt
Typeset by Six Red Marbles
Printed and bound in Latvia by Livonia Print

The authorized representative in the EEA is Penguin Random House Ireland,
Morrison Chambers, 32 Nassau Street, Dublin D02 YH68

A CIP catalogue record for this book is available from the British Library

ISBN: 978-0-241-67470-3

Penguin Random House is committed to a sustainable future
for our business, our readers and our planet. This book is made from
Forest Stewardship Council® certified paper.

Penguin Random House is committed to a
sustainable future for our business, our readers
and our planet. This book is made from Forest
Stewardship Council® certified paper.